With very best wishes,

Philip Evans.

D1064603

BEDFORD SCHOOL
AND THE GREAT FIRE

BEDFORD SCHOOL
AND THE GREAT FIRE

M.E. Barlen, M.P. Stambach and D.P.C. Stileman

Quiller Press
London

Frontispiece *(Bedford County Press)*

First published by
Quiller Press Ltd.,
50 Albemarle Street,
London W1X 4BD

Copyright © 1984 Bedford School

All rights reserved. No part of this book
may be reproduced or transmitted, in any
form or by any means, without
permission from the Publishers.

Typeset by Millford Reprographics Int. Ltd.
Designed by Peter Hurst
Printed by Burlington Press (Cambridge) Ltd., Foxton
Cambridge CB2 6SW
Design and production in association with
Book Production Consultants, Cambridge

Contents

Foreword

By the night of Saturday, 3/4 March 1979, the Main Building of Bedford School had existed for over eighty-seven years and a span of seven headmasters.

It is strange that it is now apparently so easy to look back over five years to that night. An evening which had begun with a French play in the Memorial Library and had continued until just before midnight at an enjoyable social gathering ended in apparent catastrophe. The frightening glow over the town we suddenly noticed on our drive back to Bedford took us into the School Estate only minutes after the fire engines had arrived.

The first enquiry as to what had happened, directed at the Bursar, who also had only just arrived was, I remember, monosyllabic. Then suddenly, as we were joined by others, individual comments helped to focus the attention and concentrate the mind on what was happening. These were the first of many uplifting experiences. A housemaster quite simply pledged his support; an Old Bedfordian on the point of retirement expressed his sympathy; a senior master commented on the qualities now required; and a senior boy put the facetious comments of a bystander in perspective. After what seemed an age the firemen allowed us into the building, and two of us dropped the offset litho down the stairs! As it turned out we could have left it where it was and ironically it would have remained intact!

For many watching the spectacle, though, there was an acute sense of helplessness, and the real extent and impact of this only emerged later when the boys expressed their own feelings and reactions to what had happened. They were, more importantly, both what we would have hoped for and expected.

The next morning, Sunday, 4 March, emphasised the reality of the situation – it could never happen to us, but it had. After the total destruction of the Main Building thirty classrooms had to be found from somewhere, plus a masters' common room, and the problem of where to stage everything that normally took place in the Great Hall had to be solved. The immediate

response by everyone was magnificent and remained so. People simply got on with whatever they were asked to do. Looking back there was a remarkable absence of recrimination. How after all do you explain to your Governors that the Main Building of the School for which you are responsible has burnt down? 'Never mind,' said one very distinguished Old Bedfordian Governor, 'buildings like that should only last about sixty years.' Strangely, the cause of the fire seemed to be the last thing on everyone's mind.

Far more important was dealing with the problems at hand. School had to continue with the minimum of disruption and plans had to be made for the immediate and long-term future. Consequently the most positive effect of the fire was to emphasise to all of us that life is about people. And the people, boys, staff, wives, parents, Old Bedfordians, Governors and friends of the School all responded in a variety of special ways.

The restored building viewed from the school field. *(Morley Smith)*

Boys, realising that they were unlikely to be spoon-fed, started to solve problems they would previously have happily handed on to someone else. Embryonic barrack-room lawyers were put in perspective, and leadership qualities emerged which had previously been unrecognised. Full marks to the ebullient fifth former who told a television interviewer that he had not lost any notes in the fire because he had taken them home to revise over the weekend. His eventual 'O' level results proved the point as well! Thanks to that sort of independence of outlook and to the optimism, cheerfulness and helpfulness of every single member of the School we succeeded in our short-term objective which was to complete the term with the minimum of disruption.

During this time a reorganised staff structure evolved. Schoolmasters suddenly assumed the roles of architects, designers, planners, public relations and security officers, and above all became experts in improvisation. Nobody stopped to work out how many man-hours of academic work and preparation in terms of teaching notes accumulated or how many boys' revision notes for Ordinary and Advanced Level examinations less than fourteen weeks away had been lost. Instead, individuals set about their own tasks, sometimes on their own, sometimes with the help of others. Parental typewriters worked into the small hours, the results were quickly duplicated and many boys ended up better off than before! Furniture arrived for the School Office, a duplicator to help the Common Room, a secretary seconded to help with all the insurance claims, the Local Education Authority offered classroom furniture, and other offers of assistance arrived with each post.

Governors somehow made time to become even more committed to the needs of the School, and above all to take action and make crucial decisions whenever they were necessary. One decision in particular, to continue with the building of the Recreation Centre, was approved unanimously as a result of a masterly interjection in committee which cut short any chance of a pessimistic debate. After this any doubts or reservations were minimal, the main object being quite simply to see the School returned to normality in an improved state with the target date September 1981.

With so many people concerned and involved, the School really did become the whole community which in its idealistic moments it had set itself out to be. As a consequence of this the School's horizons widened; after the television crews and reporters came the consulting engineers, Austin Hall came to supply the temporary classrooms, and Phoenix to answer and settle the vast number of insurance claims large and small, which had been collated and coordinated by the Bursar with the efficiency of a military machine.

The most crucial decision affecting the restoration of the Main Building was the choice of the architect. A discussion about the possibilities; a phone call by the Chairman of the Harpur Trust to Rome; the strongest possible recommendation from an Old Bedfordian, Bernard Feilden, and within days Philip Dowson of Arup Associates had appeared to look at the gutted building. Immediately we were struck by his sensitivity, sincerity and feeling for our predicament. In front of the Governors his talk was about education and people, not about architectural technicalities; the other possible architect mentioned neither. So we had our architect and this began the start of a remarkable partnership between Arup and our own Staff Planning Committee, which provided the foundation of the Restoration.

Later Arup were joined by John Laing on a management contract for the construction of the building, and we saw at first hand the construction industry at work. Initially little positive seemed to happen, lorry loads of debris left, but the mud and dirt remained, the large horizontal crane came and stayed, morale suffered and we sacked our caterers!

Then the process of building began and after a long year since the fire it became possible whenever in the building to look forward rather than to look back. The Commemorative Stone Ceremony gave the members of the School their first real opportunity to see what was happening inside the building. The Topping Out, expertly handled by our public relations adviser, Dewe Rogerson, proved to those outside the School community that progress was indeed being made.

The 1981 School holidays saw the School's stamina undiminished as, working within very tight deadlines, the departments and administration moved back into the building in time for the new School year.

The first Assembly back in the building for those in the School who had seen it all through, was an emotional experience. The Service of Dedication later in the term and the events of Restoration Year emphasised our thanks and great gratitude to everyone who had so willingly helped and supported us. In particular we as a School, past, present and future, now thank our architects for their vision and imagination; our builders for the quality of their craftsmanship; our financial advisers for their optimism and confidence in our future; and our Governors, parents and friends for their loyalty and constant support.

This book tells you something more of what we believe is a rather unusual story.

C.I.M. JONES
Head Master
Bedford School
1984

Chapter I

Bedford School 1892–1979

The story of Bedford School before the fire begins properly with the opening of the School Building in 1891, and with James Surtees Phillpotts who built it. Phillpotts, firmly fixed by the dates of his Headmastership in the last quarter of the Victorian era, was essentially a twentieth-century head-master, for it was he who wrested the School from local obscurity in the nineteenth century into national significance in the twentieth. To say that he built the New Schools, as the building was called, is not a simplification. Of course others worked at the project. There was an architect to design it, Governors to debate and approve the plans, a builder to put the fabric together; but without Phillpotts no new building would have been built.

In 1874 JSP had taken over a school of 270 pupils, in 1884 there were 540, and by 1891 the School roll had grown to nearly 800. In 1886 the Cowper Buildings had been opened (they still stand in the south-west corner of St Paul's Square) and with this extension the School had a further nine classrooms at its disposal. But even before they were opened Phillpotts had reported to the Governors of the Harpur Trust that the total accommodation would still be inadequate. The Governors must have wondered what they had taken on when they appointed this whirlwind of a man.

The details of the debate over the proposed move of the School to more adequate premises and the history of the purchases in successive stages of land between De Parys Avenue, Pemberley Avenue and Kimbolton Road are well recorded in full in Sargeaunt and Hockliffe's *A History of Bedford School*. Two facts stand out that need to be mentioned here as examples of Phillpotts's generosity and single-mindedness. He bought out of his own pocket two parcels of land that now form part of the School Estate, so convinced was he of the inevitability of the move. This land he later sold at cost to the Governors when they finally recognised the wisdom of his foresight. Second, when the finance for the new building was in doubt and the Governors refused to sign a contract until the shortfall was secured, Phillpotts sent out an appeal for funds and personally contributed £500. What this sum represents in modern terms one can safely leave to economists to evaluate.

The choice of an architect proved to be difficult. Two sets of plans were submitted in 1887 under pseudonyms and they can still be inspected at the County Records Office. One of the designs proposed a building in the shape of an open-sided rectangle, some 80 by 50 yards in dimension, with an assembly hall, placed inside the rectangle and at right angles to the longer elevation, 120 feet in length and 54 feet wide. The centre of the front elevation was designed with a clock tower surmounted by a short spire. Neither set of plans was accepted, and the committee of Governors visited London to inspect schools built on the central hall principle. On the basis of their investigations E. C. Robins, FSA, was invited to submit plans, and these were accepted in 1889.

Such was the prestige of the School that both the ceremony of the foundation stone laying and the official opening aroused considerable excitement and interest amongst the general public. On both occasions the entire School, masters and boys, with Governors and Aldermen, marched from the Old School to the new site. The foundation stone was laid on 17 October 1889 by Samuel Whitbread MP, Chairman of the Governors, who had personally given £1,000 to the subscription fund. On this occasion the Mayor, Mr J. Hawkins, said: 'I know of no event which has taken place during this century of so great importance to the town of Bedford as laying the foundation stone of the New Grammar School.' The building was officially opened by the Duke of Bedford on 29 October 1891. The *Bedfordshire Times and Standard* reported the proceedings at great length.

The Opening Day, 29 October 1891. Phillpotts is seated in his chair of state which was brought up from the Old School for the occasion. It still exists. Seated to his right is the Duke of Bedford who performed the opening ceremony. On Phillpotts's left is Earl Cowper next to whom, with his face turned to the left, is Samuel Whitbread, Chairman of the Governors. The builder, T. Spencer, is seated in the front row on the left hand side.

On Monday morning, in spite of very wet weather, several men were engaged erecting scaffold poles for the triumphal arches at sundry points and in making other preparations for the decorations of the town. During the next three days displays of bunting blossomed here and there and on Thursday morning the principal streets burst forth in their full and brilliant pageantry. An exceedingly gratifying circumstance and one likely to strengthen the good feeling that had prevailed between the town and the School was the cordial invitation issued by the Headmaster, masters and boys to the townspeople generally to participate in the rejoicings. The Grammar School field was thrown open to all and sundry, and although it was impossible for a tithe of the people to be present in the building to witness the opening ceremony they were given facilities for going over the School in the afternoon while the big people were partaking of the Mayor's and the Headmaster's hospitality. Besides what there was to be seen in the streets in the way of decorations and processions there was a football match played between the School and the Old Bedfordians, a Greek torch race with Bengal lights, and fireworks by Messrs Brock, the eminent pyrotechnics and decorative artists, provided by the Headmaster, masters and boys for the delectation of the other schools and the public at large, and a torchlight procession of cyclists in masquerade costumes.

At the end of the Easter term of 1903 JSP retired. His successor, John Edward King, had no easy task. He had to live under the shadow of a man who by general assent ranked as the second founder of Bedford School. King's headmastership was a short one, for in 1910 he moved to become Head Master of his old school, Clifton College. Perhaps some Bedfordians saw this as an act of slight disloyalty to the School, for the three previous Head Masters, Brereton, Fanshawe and Phillpotts, had stayed in office till retirement. One must therefore be all the more careful not to overlook what King did for the School.

In 1857 Fanshawe had secured permission from the Vicar and Churchwardens of St Paul's to use the Chancel for Sunday afternoon services. This facility was withdrawn in September 1903, and the School was allowed to use Holy Trinity Church by the incumbent. In October 1903, in only his second term as Head Master, King, with other signatories, sent out a letter to parents appealing for funds to erect a Chapel. At a well-attended meeting on 16 December 1903 King discussed the proposals. He revealed that the Governors of the Trust felt that the provision of a Chapel did not fall within their powers, however much they might be in sympathy with the idea.

Voluntary subscription was therefore the only remedy, and by the end of the year £2,200 had already been subscribed. The subscribers list for June 1904 records the gift of £1,000 from the Rev. T.C. Fitzpatrick. Fitzpatrick was an Old Bedfordian who had won an Open Scholarship to Christ's

James Surtees Phillpotts, Head
Master 1874–1903.

College, Cambridge, where he had taken a First in Physics. After taking holy orders he became Dean of Christ's and later President of Queens'.

In April 1906 King reported that G. F. Bodley had been invited to submit plans, and these met with the approval of the Executive Committee. Work on the foundations of the Chapel began in November 1906 when the subscription list had passed the £5,000 mark. *The Ousel* of the time pointed out that Mr Bodley had just been appointed architect of a new cathedral in Washington and that he was currently in America. The foundation stone was laid by the Lord Lieutenant, Lord St John of Bletsoe, on 18 May 1907, and the Chapel was dedicated by the Bishop of Ely on 11 July 1908. In the evening Dr Harding gave an inaugural recital on the Norman and Beard organ in aid of the organ fund, to which Phillpotts had given £106.

The cost of erecting the Chapel exceeded quite considerably the sum envisaged. Items in addition to the original contract included the sinking of the foundations down to the bedrock, the provision of steps at the west end caused by the slope of the site, and the addition of vestries at the east end which had not appeared in the original scheme. Bodley had planned that the east window should be unadorned, but the Old Bedfordians insisted that this window should be decorated with stained glass in memory of Old Bedfordians who had fallen in the South African Wars, and reluctantly Bodley designed tracery and engaged a designer for the figures of the saints which now fill it. To make good the debt of £4,000 (the voluntary subscriptions having raised £6,000) it proved necessary to incorporate a Bedford Grammar School Chapel Society as a guarantee company, and this is the origin of the Chapel Society as it exists today and of the Chapel's independence from the Governing Body of the School.

Bodley's designs for the seating in the Chapel were extant but not executed at the time of the Dedication, and a notice in *The Ousel* of 27 October 1908 was the first intimation of the plans to carry out the furnishing as it is seen today.

The next major addition to the School Estate was the Memorial Hall which was opened by Prince Henry, later Duke of Gloucester, in October 1926. It was natural that the question of a memorial to those Old Bedfordians who gave their lives in the Great War should be the province of the Old Bedfordians Club. In the minutes of the Club Committee meeting for 16 March 1915 it is recorded that a sub-committee was formed to draw up some scheme of commemorating the services of past members of the School in the War. Initially the scheme sought to raise funds to provide fee-remission for boys whose fathers had been killed, but the form that a record of those killed would take was still undecided. By 1918 Exhibitions which granted full remission of fees were offered to the sons of OBs killed in the War, but it was in a letter written on 22 February 1918 by Reginald Carter, Head Master 1910 to 1928, to the Chairman of the OB Club Committee that set plans for a visible memorial in motion. He wrote:

. . . and I feel that we should also have a visible memorial for future generations. We have no proper library and no museum, and it would be a very great help to the life of the School if this defect were remedied: also I think we ought to have a hall of moderate size in which the names of those who have fallen could be placed in a worthy setting . . . I dream of a building on the West side of the School – well designed with a fine North end – to contain a Hall and Library and Museum . . .

This proposal was unanimously accepted by the Committee and after a slow start the sum of £6,200 had been subscribed by July 1923. The minutes for the meeting of 19 July record that a committee should be set up with powers to consult an architect as to whether plans for the proposed building could be carried out with this sum. It is here that the name of Oswald Milne first becomes linked with the School Estate. It is not clear why he should have been approached. He was an Old Bedfordian, and it is the view of A. D. Nightall, the present Secretary of the OB Club, that he was chosen simply because he was known personally to the Committee. For some this choice might be seen as a blatant example of Old Boy parochialism, but as events proved it was well nigh a stroke of genius, for Milne became with this first commission the Architect to the School, and the appearance of the School Estate as it was at the time of the fire was almost entirely due to him.

On 24 June 1925 Milne's plans for the Memorial Hall and Library as it stands today were approved by the Planning Department of the Local Authority, and with the consent of the Estates Committee of the Harpur Trust building work began. The builder whose tender was accepted was Messrs Samuel Foster of Kempston, the builder of the Chapel, and it was this local firm that was to act in partnership with Milne on all of his designs for the School.

The foundation stone was laid on 26 July 1925 by Lieutenant-General Sir Walter Braithwaite KCB (OB). The weather was appalling on the day and proceedings were delayed in the hope that the rain might ease off. It did not, and spectators took what shelter they could find under umbrellas and neighbouring trees as the dignitaries and a Guard of Honour formed from the School OTC took up their positions. Under the foundation stone was placed a bottle containing newspapers, *The Ousel*, and other documents of interest.

Exactly one year later the Memorial Hall was opened by HRH Prince Henry. A Guard of Honour lined Burnaby Road, and the Prince was welcomed by Sir Maurice Craig CBE (OB), the President of the Old Bedfordians Club which had financed the entire building project. In his speech Sir Maurice reminded the gathering of the record of Old Bedfordian service in the War. Of the 2,350 members who had served in the armed forces, 454 had been killed, and they were remembered in this Hall, their names being carved in gold lettering in ten blue panels on the north wall.

After the Opening the building was formally presented to the Harpur Trust by General Sir Walter Braithwaite.

(Twenty-three years later, on 12 June 1949, panels at the south end of the Hall were unveiled in memory of 274 Old Bedfordians who had fallen in the Second World War. Again this memorial was the gift of the Old Bedfordians Club, and it was unveiled by General Sir Sidney Kirkman, Quartermaster-General to the Forces.)

It is at this point in the history of the School, that is in 1926, that the name Eugene Rolfe begins to take on the great significance it was ultimately to bear. Rolfe was Librarian at the time of the opening of the Memorial Library, and it was he whose energy ensured that the plans for the furnishing of the Library were executed. The designs for the book bays, the chairs and the tables were the work of Oswald Milne, whose artistic sense did not stop with the design of the building but was expressed in the minutest detail of furnishing. Rolfe must count as one of the most devoted members of Staff the School was ever lucky enough to possess, for he pursued his goal to have Milne's plan realised with exceptional determination. The School Archives contains box-file after box-file of correspondence relating to the appeal for furniture, and each issue of *The Ousel* from 1927 publishes Rolfe's reports on the progress of the appeal. The idea of a personal gift bearing the donor's name engraved on a bronze plate found rapid favour, despite the enormous cost. Armchairs costing £3 10s., smaller chairs costing £2 10s. and tables at upwards of £14 were not cheap, but the permanence of the finest oak was a great attraction.

By March 1928 the equipment of the Library was complete, but Rolfe was not the man to allow this wave of generosity to lose momentum, and he immediately turned to the furnishing of the Memorial Hall itself. Friends and past members of the School were invited to donate 144 chairs and 2 tables, again to the designs of Milne, and progress on this venture was well under way when Reginald Carter retired at the end of the Easter term 1928.

Humfrey Grose-Hodge was only thirty-six years old when he became Head Master in the Summer term of 1928. One of his first official functions was to attend the annual Dinner of the Old Bedfordians Club at the Trocadero Restaurant on 31 May. A list of those Old Boys attending shows that at least two-thirds of them were older than he, and at least a third had left the School before he was born. But, thanks to the strong links that Carter had forged with the Club, Grose-Hodge knew that he had an enormous body of support on which he could draw for his plans for the School. It was therefore with a certain self-consciousness of his comparative youthfulness but with full confidence in his position that Grose-Hodge rose to address his audience on 30 July at Speech Day.

Ladies and Gentlemen, I stand before you as the new Head Master of Bedford School with a degree of pleasure and of pride, which I make no attempt to conceal. Probably many of you are thinking, as some of you

The Great Hall as originally built. The ornate oak screen at the west end was resited at the east end when the more modern panelling (see below) was installed in 1934. The bust is of Sir Erskine May (OB), Clerk to the House of Commons. At the extreme right of the picture is the memorial to Old Boys killed in the South African Wars. This plaque was covered over when the Hall was panelled in the 1930s and only revealed again after the fire.

The Great Hall in the 1960s. The Oak panelling is complete. The ground-floor south wall windows which gave borrowed light to the classrooms have been covered by panelling. The portraits of Head Masters, given by the Old Bedfordians, included Brereton, Fanshawe and Phillpotts (out of sight) then King, Carter, Grose-Hodge and Seaman. Brown's portrait was hung in 1972 on the north wall. Note the platform furniture, all the gift of Old Boys. On the north wall the panel commemorates the visit of HRH Princess Margaret in 1952. Note the balustrade; the vertical bars were black, the scrolls gilt, the handrail was made of oak. The Latin summary of the School's history is seen below the balustrade, gold lettering on a black ground. The roof, made throughout of pitchpine, is perhaps the most impressive feature of this fine architectural achievement. The beams with hammer ends were solid timber. Note the repeated quatrefoil motif in the braces of the beams and the wall facings. Note also the light from the dormer windows which lit this area of the Hall and externally relieved the monotony of the enormous roof area.

have been saying to me in the last day or two, that I look very young. I assure you that I look younger than I am, or than at this moment I am feeling. But, if I am young, it is not my fault; and if it be a fault, it is one that time will very soon cure. For the moment I invite you to look on me rather as a well-known furnishing-house invites us to look upon their furniture – not as a modern fake, but as one of the antiques of the future.

Grose-Hodge referred later on in his speech to the events of the day before when, after Morning Chapel, the bronze statue of St George which stands in the niche on the north gable end of the Memorial Hall had been unveiled by General Sir Walter Braithwaite. The sculptor was Herbert W. Palliser, and with this addition the War Memorial Hall was complete. Grose-Hodge took this opportunity to link the generosity of the Old Bedfordians in their gifts of the statue of St George and of the portrait of Mr Carter with the recently completed furnishing of the Memorial Hall. He then made it plain that he hoped that a similar project might be undertaken to re-equip the Great Hall with oak chairs.

This was a challenge that could not fail to fire Rolfe's imagination. In an article in *The Ousel* for 28 July 1928 headed 'The School Hall' he refers to the scheme for giving a 'leaving chair' as a visible record of a boy's connection with the School, and he continues: 'Is the writer too sanguine a visionary when he foresees a Hall fully panelled, with the names of all Heads of School visibly recorded somewhere in it?'

His vision took in fact more than twelve years to complete, and even though he retired from the Staff in 1934 after thirty-seven years of service (he had been appointed by Phillpotts) he continued to work unremittingly at the cause with which he so completely identified himself. The reader who cares to study the bulletins that he wrote for each issue of *The Ousel* (from 1931 the magazine appeared twice-termly), initially under the heading 'The School Hall' but from 1929 onwards under the heading 'Oak Fund', will readily gain the impression that for Rolfe the adornment of the Hall in oak became an obsession. Repeatedly he reminded his readers of the progress so far achieved and exhorted them to further acts of generosity. At times his notes have a whimsical touch: referring to the possibility of displaying the coats of arms of New College and of the Town of Bedford he writes: 'Dare one venture to hope that the bodies which govern these venerable institutions will – parentally, one might say – supply the wherewithal for making possible these permanent and public records of their ancient connection with us?' But such lapses, if not rare, are entirely forgivable when one considers what he achieved by his enthusiasm and persuasiveness.

Important stages in the adornment of the Great Hall, as it came to be known, were the gift of the three chairs of state to the design of O. P. Milne in June 1929, and the completion of the panelling of the south wall of the ground floor. February 1931 saw the start of the long-term project to replace the pitchpine railings with a wrought-iron balustrade, again to Milne's design, which was built in the School Workshops. In June 1931 the School Arms, for which Milne himself made a gift of £98, were placed above the Head Master's chair at the west end of the Great Hall. The clock was unveiled on 18 November 1933 after the Haileybury match (won 15–5). This took the form of an octagon one yard across of ebonised wood with gold hands and figures with the adage VT HORA SIC VITA. During the Christmas holidays 1933–4 the panelling on the north wall was completed, and coinciding with Rolfe's retirement at the end of the Summer term of 1934 came the news that Sir Richard Wells MP (OB) had donated funds for the building of the ground-floor west screen to replace the ornately carved screen that had stood there since the Opening in 1891. One would like to think that the timing of this very generous gift was not a mere coincidence. 2 March 1935 saw the unveiling of the shields emblazoned with the arms of the Town of Bedford and of New College. The arms of the Town were given in memory of R. J. A. Dawes (OB 1889–93) by his widow and his son, A. G. Dawes (OB 1918–27), and the arms of New College, given by the College, were presented by the then Warden, H. A. L. Fisher, the historian. During the summer holidays of 1938, by which time the panelling throughout the Great Hall had been completed, an inscription in gold lettering on a black gold-bordered ground was placed on the lower part of the Middle Gallery along the east, south and west sides of the Hall. The inscription runs:

SCHOLA BEDFORDIENSIS HAC IN VILLA ANTIQVITVS A MONACHIS INSTITVTA CHARTA A REGE EDVARDO VI° ANNO DOMINI MDLII DONATA A GVILLIELMO HARPER EQVITE ET DOMINA ALICIA VXORE EIVS DOTATA DEINDE ET INSTAVRATA TRIBVS EXACTIS SAECVLIS IN AMPLIORES HAS SEDES TRANSLATA NOVIS POSTEA AEDIFICIIS ORNATA AVCTA DEO DVCE FLOREAT

The translation runs: 'Bedford School, founded in this town by monks in ancient times, granted a charter by King Edward VI in the year of Our Lord 1552, endowed and renewed thereafter by Sir William Harper, Knight, and Dame Alice his wife, transferred after the lapse of three centuries to its more spacious home, afterwards enlarged and adorned with new buildings. May it flourish under the guidance of God.'

The outbreak of war saw a severe curtailment in the work of the Oak Fund, but gifts were invested in Defence Bonds in anticipation of peacetime, and indeed sufficient money was contributed to complete the panelling of both ground-floor corridors that led into the Great Hall. Sadly enough, Rolfe died on 20 April 1942, before this work was completed. Few Old Bedfordians can remember the Great Hall as it originally was. The majority carry memories of the august splendour it achieved thanks to the unyielding efforts of Eugene Rolfe. It is difficult to imagine what would have been his reaction had he known that, barely forty years after the bulk of the work had been achieved, the whole edifice was to be destroyed by a single act of hideous vandalism.

Against the backdrop of this steady labour of adornment there started a phase of building without parallel in the history of the School Estate. Grose-Hodge was resolved to give Bedford School the amenities that would place it firmly in the first rank of English public schools. The School enjoyed a full range of facilities, but in many cases the housing for them was outdated and in some cases an eyesore. Grose-Hodge took up the ambitious vision that the 'Old Chief', as Phillpotts was affectionately known, had had, and gave it his personal impetus.

In December 1929 Grose-Hodge sent a letter to all parents inviting subscriptions towards the cost of building a swimming bath. Concern about overcrowding and health risks attached to the use of the Corporation Baths next to Prebend Street Bridge had made the need for the School to possess its own pool an urgent one. O. P. Milne's provisional plans were published in February 1930, though they were later amended to include housing on the north-east corner for the chlorination plant. The Pool was opened for use on 16 June 1930, barely four months after the site had been pegged out. It was an attractive feature of the School, built of mellow brown-red brick to harmonise with its near neighbour the Memorial Hall. Enclosed by a ten-foot wall skirted with lawns, it housed a pool some 100 feet by 40, with a masonry diving stage of four levels. At the shallow end stood a fountain, and outside the Pool on the north wall a notice board proclaimed from day to day the temperature of the water. For over forty years it served the needs of the School, and several thousand boys must have learned to swim in it and to swim the 'pass', a seventy-five-yard swimming test, which gave them the right to remove from their caps the little white button, the tell-tale sign of a non-swimmer at Bedford School.

Many will have mourned the demolition of the Swimming Pool when the plans for the Recreation Centre were approved, but it must be conceded that the Pool was often unusable before half term in the summer, and that top-class swimmers are reared nowadays in indoor pools. It is of interest to note in passing that then, as now, the use of the Pool was extended to outsiders, in 1930 to the Physical Training College and the High School. The present writer has a clear memory of an unwritten Lower School rule which forbad boys to look out of the top floor windows during morning school in the Summer term lest the sight of girls bathing should kill their appetite for lessons.

In the same year that the Pool was opened the planning for the creation of some tangible record of the affection in which Phillpotts was held by Old Bedfordians was under way. A letter sent to all members of the Club in March 1929 by a special committee had indicated the choices. These were

The Swimming Pool seen from the south west corner. At the far end is the fountain and to its right in the corner is housed the chlorination plant. The site is now covered by the Sports Hall.

the foundation of a scholarship to bear his name, the erection of ornamental gates at the Burnaby Road entrance, and the completion of the panelling in the Great Hall. The 'Old Chief' had been consulted as to his views and the letters that he wrote to Sir Walter Braithwaite at the time are characteristic of the great man. A scholarship was in his opinion far too costly in terms of the capital investment required, and he deprecated any attempt to raise a sum of money that might in any way penalise the resources of the donors. It was therefore agreed to erect the Gates, for the panelling was well in hand and could be seen as a long-term project. Sir Edwin Lutyens was consulted over the appointment of an architect and he recommended his former assistant, O. P. Milne. This could not fail to be a gratifying choice for the Club since Milne was already so closely linked with the new work on the School Estate.

On 18 July 1930 Phillpotts was ninety-one, and on 8 October he was, for the first time in twenty years, too ill to attend his annual 'At Home' to the Old Bedfordians of his era. His son Geoffrey stood in for him and read the message that JSP had sent:

'My colleagues and Old Boys of Bedford School, I am very sorry that for the first time I am unable to be with you in person for our annual tea.

21

I grieve to miss it, as it is our twentieth gathering today. It is a great deprivation to me, as I have always looked forward with the greatest pleasure to this yearly opportunity of seeing so many of my old friends, but, alas, my doctor has forbidden it.

Among my many acceptances for this afternoon was one from Air Marshal Sir Sefton Brancker, but today, not only all Bedfordians, but the whole country mourns his loss. FLOREAT BEDFORDIA.'

(Sir Sefton Brancker, KCB, AFC [OB 1891–94] was Director of Civil Aviation and was one of the victims of the R101 disaster.)

The Gates were opened by Sir Reginald Lane Poole on 18 October 1930 before a distinguished audience, but the ceremony was overcast by the sad fact that Phillpotts had died two days previously at his home. The ceremony was immediately followed by a Memorial Service in the Chapel at which the address was given by the Rev. T. C. Fitzpatrick, DD, President of Queens' College, Cambridge (OB 1869–80). The music chosen for the organ voluntary was the prelude to Elgar's *Dream of Gerontius*.

This, then, was the end of the most important figure in the modern history of Bedford School. Tributes published in *The Ousel* at the time occupied much space and included the obituary printed in *The Times*. One article gives great insight into the man. It is an extract from a letter written by Phillpotts to the then Head Master of Bradfield which the latter sent to the Editor of *The Times*. Phillpotts had written:

When I went to Bedford in 1874 Jex Blake said, 'Remember you will never be again the light-hearted man you have been here (Rugby), for you will always have a heavy responsibility hanging over you.' I had to get rid of the thinking of the responsibility in order to keep my brain clear, and the only way I could do it was getting on my horse and jumping over hurdles or fences, because, if I did not give my whole mind to the jumping, I was sure to come a cropper. Ordinary riding was not enough.

At Speech Day 1931 Grose-Hodge referred to plans to extend the Preparatory School Buildings. The original building had been erected in 1899 on the site of the old St Peter's Rectory, thus providing a permanent home for the Preparatory Department. It is interesting to note that this site had been considered for the Main School Building but had been rejected on the grounds that there was insufficient space between the graveyard of St Peter's church and the houses to the east in St Peter's Street.

The partnership of O. P. Milne and Samuel Foster Ltd which had proved so successful for the Memorial Hall and the Swimming Pool was employed for the new extension, and it was opened on 13 May 1932 by the Head Master in a very simple ceremony. The extension harmonises well with its parent building and the roofline and the window details are interesting in that they anticipate the basic features of the Science Block, which was to be Milne's next project.

Old Bedfordians whose link with the School goes back to the early thirties will recall the 'Tin Buildings'. These were in fact one building that stood in front of the Fives Courts and Workshops at the south end of what is now the main Inky level. It was classed as a temporary building by the Local Planning Department, and the County Records Office contains details of planning permission granted every ten years for the continued use of the structure from 1889 to 1929. Originally intended to house forms for which the Old School had no room, they were retained when the New School was opened in 1892 and were converted into a gymnasium at one end and science laboratories at the other. Phillpotts had introduced the teaching of Chemistry and Mechanical Science to the School in 1875 and appointed Mr A. Talbot (after whom the Boarding House is named) to teach these subjects. Interestingly enough, this innovation was taken up by the Modern School whose Upper Forms attended Talbot's lessons side by side with the Grammar School boys in the early years of Phillpotts's reign.

By the time Talbot retired in 1915 Science formed a major part of the School curriculum and its teaching was centred in the Tin Buildings. But with the passage of time the Buildings were becoming outdated, and, in Grose-Hodge's words at Speech Day 1932, they were 'singularly unlovely'. And so plans for a Science Building were put in motion.

No doubt practicalities such as easy access to water, gas and electricity services played their part in the siting of the building, but there can be little doubt that Milne wanted to balance the position of the Memorial Hall and to provide a fine vista of the School Estate for the visitor who approached via Burnaby Road. At that time the visitor's first impression was of the side of the houses in Glebe Road and some inadequate gates. Milne intended that in future the massive gable ends of his design should form an impressive background to the eastern side of the playing field. Part of the plans included the building of the Glebe Road gates as we see them today, and it is worth commenting that for the design of these gates he chose a style less ornate than that of the Phillpotts Gates and more in keeping with his design for the Science Building.

The Regius Professor of Physic at Cambridge, Dr W. Langdon Brown (OB 1879–88), laid the foundation stone on 29 October 1932. In the speeches that were given, Grose-Hodge felt sufficiently emboldened to describe the old Tin Buildings as 'undeniably hideous', and Dr Langdon Brown paid tribute to Mr Talbot who was present in the audience (he died in 1940, aged 88) and said that the School owed an enormous debt to his work.

The removal of the Tin Buildings could only be carried out if the Gymnasium also were to be rehoused. Grose-Hodge's extraordinary energy and persuasiveness saw to it that no time was to be lost, and in *The Ousel* of February 1933 Rolfe described in a characteristically enthusiastic article the outlines of what Milne proposed. The new Gymnasium with a ground floor of 80 feet by 40 feet and a large gallery with office and changing rooms beneath was to be erected to the immediate south of the Swimming Pool and thus form a continuation of the line of the School's western boundary. Work began early in 1933 and proceeded concurrently with the work on the Science Building. Both buildings were taken into use at the beginning of the Christmas term of 1933. Again, one is forced to admire the speed and efficiency of the builder of both projects, Messrs Samuel Foster of Kempston.

The impact that the first five years of Grose-Hodge's Headmastership had had on the appearance and the amenities of the School is difficult to exaggerate. Since his arrival in May 1928 he had extended the Preparatory School building, built the Swimming Pool, removed the eyesore of the Tin Building and erected a Gymnasium and a Science Block, whilst the Old Bedfordians had erected the Phillpotts Gates and made a very confident start to the adornment of the Great Hall with oak panelling and furnishings. The effect of these changes must have amazed Old Bedfordians returning to the School. From a practical point of view the School was able to offer vastly improved facilities that surely matched, if not surpassed, those of schools of equal reputation. From an architectural point of view, the Estate was enormously enhanced. Untidy corners had been cleared away and the west and south sides of the Estate had taken on an appealing harmoniousness of architectural beauty. The roadway leading through from the Main Building to the St Peter's entrance had been realigned and birch trees had been planted. All this was due to the dynamism of Grose-Hodge and the sensitive

View of the gymnasium, left, next to the Swimming Pool. *(Morley Smith)*

eye of O. P. Milne. Such was the enthusiasm for Milne's work that some felt it was a pity that he could not be commissioned to rebuild the main teaching block. Taste in architectural fashion changes quickly, and it is not surprising that for a period the design of our Main Building found disfavour (perhaps it still does) amongst some for its Victorian interpretation of Tudor architecture on a vast scale.

None of these additions had, however, been officially opened by some eminent dignitary, and so it was a rare coup for Grose-Hodge that he was able successfully to invite the Prince of Wales to visit the School in November 1933. The Prince's brother had come to open the Memorial Hall only seven years before and it was with a justifiable sense of pride and achievement that the School welcomed a second Royal visitor to see the changes wrought in a very brief span.

The excitement of the day was somewhat shaken by the delay in the Prince's arrival. The Guard of Honour had been brought to attention at 11.20 for the scheduled arrival at 11.30, but it was not till 11.45 that a telephone message announced that the Prince of Wales had just landed at Henlow. However, all passed off well. The Prince visited the Science Block, observed classes in action and then moved down to the Biology Laboratories, which were housed in those days over the Workshops on the southern side of the school. Then followed an inspection of the Preparatory School, the Gymnasium and the Swimming Pool, where the Prince commented on the flatness of the water, and finally he was taken to see the Library and the Memorial Hall, where he signed his name in the Visitors'

Book beneath that of his brother. By this time the School had assembled in the Great Hall where they were addressed by the Prince. He referred to the long history of the School and to a number of Old Bedfordians whom he personally knew. He then announced that he had asked the Head Master to grant an extra five days' Christmas holiday which elicited considerable cheering.

The Pavilion, which had been opened on the same day in 1899 as the Preparatory School, had by 1934 proved itself inadequate in size for the uses it was now being called on to fulfil. Total rebuilding was fortunately unnecessary, for Milne was able to use the existing walls and foundations and to extend these into the building we know today. Finance for the project seems to have been a tricky question. Grose-Hodge evidently felt that, given the financial climate of the times, it was wiser to try to repeat the successful fund-raising bazaar of the twenties when £3,000 had been raised for the Boat Houses. Grose-Hodge accordingly wrote a letter to the local press, advertising the bazaar and urging public support. His letter shows how aware he was that the townsfolk might frown on the School for undertaking yet another building project and calling for yet more money in a 'period of unparalleled depression in Bedford'. But he argued his case well, pointing out that all the building schemes that the School had undertaken in the last four years had provided employment for workers in the building and allied trades. His letter must have had its intended effect, for on 14 and 15 June 1934 all but £14 of the £1,500 needed was raised. Stalls were set up in the Great Hall, and a one-act farce was given a twice-daily performance in the Gymnasium, as was a 'variety entertainment'. There were swimming displays, an amusement park on the Inky Field and pony and donkey rides. The Bazaar was also honoured on the second day by a Royal visit. Coming so close on the visit of the Prince of Wales, this was a surefire attraction. The Prince Sayid-Bin-Matata of Abyssinia, with a retinue of four, was welcomed

The School Fete, organised to raise funds for extensions to the Pavilion, was visited by an Abyssinian prince and retinue. This hoax, into which Grose-Hodge had been initiated in advance, passed off without discovery. Note the chairs of state. The photograph was kindly lent by E. T. L. Spratt.

by the Head Master, via an interpreter, on the platform of the Great Hall before a packed audience. Speeches were exchanged, there was due applause, and the Prince and his robed retinue inspected some of the side-shows; but when they sensed that their greasepaint was beginning to melt, they left in their hired Rolls-Royce. Grose-Hodge had been initiated into this hoax, which is what it all was, well in advance, and had played along with the four Old Bedfordian Cambridge undergraduates. He even went as far as posing in his chair of state with the hoaxers for the press, and the resulting photograph was published together with many others the following Tuesday in the *Bedford Record.* It is hard to believe that the hoax was not recognised on the day, but the terse editorial comment of the *Bedford Record* suggests that it was totally successful.

On 6 January 1935 Sir Maurice Craig, CBE, MD, FRCP, died. Craig had been President of the Old Bedfordians Club at the time of the opening of the Memorial Hall, but more significantly he had been Chairman of the Harpur Trust from 1929 to 1934 when Grose-Hodge's plans for the development of the School Estate were in the process of fulfilment. In his professional capacity he was a leading authority on the treatment of nervous disorders. He is remembered in the School in two ways. His widow presented to the school the two Thompson tables which by great good fortune survived the Fire. Originally they were placed one at the west end, the other at the east end of the School. Carved from massive slabs of oak, they bear the maker's trademark, a fieldmouse, carved on the pedestal. His second memorial is of course the Craig Building. Grose-Hodge was well aware of the inadequacy of provision for the needs of boys who lived at a distance from the School. At Speech Day 1936 he announced that a new building was to be added to the south of the Howard Building to provide further changing and recreation accommodation for dayboarders. Milne undertook the design, and the Craig Building was opened for use in the Easter term of 1937.

The years of the Second World War saw no building development, unless one thinks of the brickwork and sandbagging in front of the ground-floor windows on the south side of the Main Building which were put up to create refuge rooms in case of air attack. But one incident stands out from these years. On 28 February 1945 the roof of the south side of the School caught fire, and by the time the fire brigade were directing their hoses on the blaze the beams were burning from end to end. *The Ousel* reported:

> It seemed only a matter of minutes before the flames would spread to the roof of the Great Hall; and if that happened, nothing could save the School. The dry pitch-pine of the roof would go up like a torch, till the 'copper spike' crashed down, to hurl fragments among the chairs and to fire the panelling of the walls.

Some vision! But the fire was put out by nine in the evening, two hours after its discovery. Water damage was immense, but this was a small price to pay.

27

Morning Prayers were conducted as usual, thanks to a band of twenty boys who had straightened up the mess in the Great Hall, and the next three days were spent clearing classrooms of damaged furniture and sodden debris. Sawdust was strewn to mop up the water and with all the windows thrown open the sun and wind of perfect March weather dried the building out. Letters of concern and sympathy arrived from the School's many friends, and one Old Bedfordian wrote: 'I would have walked from London with a bucket of water if it would have done any good'.

In 1952 the School celebrated the 400th anniversary of its Charter granted in Letters Patent by King Edward VI. The celebrations took the form of services in St Paul's Church in June and July, at which the Bishop of Peterborough and the Bishop of Blackburn preached, but the main event was the visit of HRH Princess Margaret on Saturday, 14 June, on the occasion of Speech Day and Prize Giving. This was C. M. E. Seaman's first year as Head Master. Among the many guests was Humfrey Grose-Hodge whose Head Mastership had brought so much distinction to the School, and to whose service Mr Seaman paid a most generous tribute in his speech. The commemorative panel which Princess Margaret unveiled over the north door of the Great Hall had been designed to match the Latin inscription on the Middle Gallery, being gold lettering on a black ground. It was the gift of O. P. Milne.

Under William Brown, Head Master 1955 to 1975, three major developments took place. The first of these was the Hayward Wells Building which allowed an extension of the Workshops and the provision of a separate Geography Department. To this end the Forge had to be sacrificed, a sad day for Old Bedfordians who remember amongst others Mr Salisbury, the doyen of the Forge, whose excellent craftsmanship created the Great Hall balustrades and the lighting chandeliers in the Chapel. Instruction in carpentry and metalwork has a long history at Bedford School. They had been introduced as subjects for practical study by Phillpotts in 1876, somewhat to the disgust of some parents, and even saddlery was taught at the School in the latter years of the last century.

No permanent site for these activities could be found until 1895 when the building currently in use was erected at the south end of the Estate. The School Prospectus for 1911 refers to the facilities which included steam and gas engines, dynamos, accumulators, lathes for wood and metalwork, forges, as well as drilling and milling machines. Tuition was given by four full-time instructors. However, the enormous expense of re-equipping the Workshops as well as the difficulty of obtaining good machinery so soon after a major war had led to their falling away from the eminent position they had held in the first part of the century. When the decision to update them came in 1958 it was long overdue. The teaching staff were still relying on the noisy overhead belt system for driving the machines.

At the same time as plans were being drawn up for these much-needed extensions it was realised that a first floor, where obviously no heavy

The workshops in 1902. This is now the carpentry shop, the machine shop having transferred to the extensions built in 1960.

machinery could be housed, might be added to provide accommodation for a new Geography Department at no great extra expense. The late Colin Reeve, Head of Geography, and David Money together planned a layout for a Department that was at that time probably the best in the country. Plans for four main classrooms with adjacent storerooms, a half roof to be used for wave-tanks and practical demonstrations, including a vulcanicity model, and an upper roof equipped as a meteorological station were drawn up. Much of the equipment was made in the Workshops beneath, which had benefited not only from their extended space but from the generous gifts of machinery from local firms. Particular attention was paid in the Geography area to natural lighting, which was achieved by providing upper windows throughout the corridors and very large windows in the classrooms. The building was designed by Geoffrey Inskip, Architect and Surveyor to the Harpur Trust, but sadly he died before the building was completed.

The Hayward Wells Building was opened on 6 May 1960 in the presence of Colonel Wells's widow, Mrs Mary Wells, and the ceremony was attended by the whole School. Bill Brown, who had vigorously supported this first building project of his Head Mastership, reminded his audience of Colonel Wells's service to the Harpur Trust and to the School in particular and of his generosity. Over the years his gifts to the School had exceeded £30,000.

In April 1965 work began on the second major project under Brown's Head Mastership. As far back as 1945 the need for some form of central Dining Hall had been recognised. The School under Phillpotts had drawn its day-pupils from within a small radius of the centre of the town, but with the years an increasing number of boys from much further afield were being attracted to the School, and despite the adaptation of the Howard and Craig Buildings the facilities for the dayboarders who had to lunch at School were proving to be less and less adequate to the task. The Boarding Houses were also facing problems. Traditionally these had always been totally independent of the School as far as catering was concerned, and this autonomy fostered a strongly cherished family spirit within each House. But in the postwar years it became more and more difficult to recruit domestic staff and an unduly heavy burden began to fall on the Housemasters' wives.

The solution was of course the building of the Dining Halls, but Brown was well aware of the resistance he would meet over the implementation of this project. The Houses would be forced to relinquish part of the independence they valued, but Brown was convinced that the loss would be outweighed by the benefits. By introducing central dining facilities he would be able more fully to integrate the dayboys into the life of the School. The division between dayboys/dayboarders/boarders would begin to fade if the School could stay together through the lunch hour. What had previously been an unoccupied part of the School day became a very fruitful period of out-of-school activities when rehearsals and society meetings could take place, with virtually all members of the School able to participate.

Sir Douglas Gordon, President of the Old Bedfordians Club and Chairman of the Harpur Trust, laid the foundation stone on 23 July 1965, and one year later the Dining Halls were opened on 28 June by Her Majesty, Queen Elizabeth, the Queen Mother. It was a signal honour to have so dearly-loved a member of the Royal Family to perform the Opening Ceremony, and the warmth of her welcome was matched by the charm that has won her affection throughout the land. The Dining Halls, designed by Deacon and Laing of Bedford and built by M. & F. O. Foster Ltd of Hitchin, were subsequently accorded an award by the Civic Award Trust.

The last of the major postwar building projects was the extension of the Science Block to house biology teaching rooms. Unlike the other sciences, biology was introduced to the curriculum in comparatively recent times. C. W. Hansel had taught some biology in the twenties, but his field was principally physics and chemistry. It was the appointment in 1929 to the Staff of J. P. Lucas as a full-time teacher of biology that led to the existence

HRH Queen Elizabeth, the Queen Mother, opening the Dining Halls on 28 June 1966. Here she inspects the Guard of Honour, escorted by Captain J. H. Davidson, member of Staff.

of a separate Biology Department. In 1933 the Science Block was opened and the Physics and Chemistry Departments moved in. Grose-Hodge offered Lucas a teaching room in it for biology, but Lucas knew that biology was going to grow far beyond the bounds it enjoyed at the time, and being an independent-minded man he refused the offer. Instead he persuaded Grose-Hodge to allow him to take over the whole of the first floor above the Workshops for his department. Grose-Hodge accepted the plan for this independent extension, and biology began to flourish.

31

Under Bill Brown biology had so far grown as to form part of the curriculum for all boys from the IInd to the Vth forms and the annual entry into the Biology Sixth amounted to some 30 to 40 boys. The Head of Department at this time was R. W. Roseveare, and he attempted to solve the problem of accommodating such numbers by extending the laboratory above the Workshops to include rooms on the first floor of the Fitzpatrick Building, but this could only be seen as a short-term solution since there was no room for a full-time laboratory steward to work in, and with rooms sited on different levels the transfer of laboratory equipment on trolleys was difficult.

In 1973 Dr Bernard Feilden OBE, a former pupil of O. P. Milne, was asked by the Governors to carry out a survey of the probable future needs of the School. As a result of his feasibility study the Governors decided that the provision of a new biology teaching block was a priority, and in due course Messrs Feilden and Mawson of Norwich were commissioned to prepare designs. Various possible sites within the School Estate were considered, but Roseveare was quite certain that the three science departments should be reunited (under one roof), after a separation of forty years, in acknowledgement of their equality.

The design and layout of the teaching laboratories and their facilities was largely the work of a team headed by Roseveare, and their plans were based on their own practical experience as well as on what they had learned from visiting a number of schools with newly designed laboratories. The ultimate design provided five teaching rooms on two floors, together with a library, an office, a preparation room and storerooms. A service lift to convey instrument trolleys between the floors was incorporated in the building, and great care was taken to design the furnishings on a flexible plan, involving movable tables and fixed 'service stations' to allow for variations and future developments in techniques.

The foundation stone was laid on 13 March 1975 by Sir George Godber, later Lord Godber, an Old Boy of the School who had studied the sciences under Carl Hansel. The Biology Wing was taken into use in January 1976, and in acknowledgement of the debt the School owes to the vision and pioneering enterprise of J. P. Lucas, who had left the Staff in 1941 and seen distinguished war service in the Far East, it bears his name, the Lucas Wing.

In conclusion, an impression, necessarily a personal one, of the School and the Great Hall before the Fire. Like the great majority of Old Bedfordians I first met it as a child. The Staff, whose service in many cases far outlasts the span of a pupil's years at school, only come to know it through adult eyes, and with the sophisticated view comes a greater detachment, an objective assessment of a building that the experienced eye can categorise, place in its architectural context, and evaluate. But the child has no such sophistication, and by the time he has grown to the age of discernment that aura of the School has laid itself on his psyche.

For me the first impression was that it was vast. I had never seen close at

hand so large a building, let alone entered one. A dull, grey March day, the war less than a year concluded, an echoing corridor, high panelling with names in blue lettering on doors, and then the Hall. A forest of chairs, a sombre gloom backed by the sunless glare of enormous windows, and height, unbelievable height, roofed in by dark timber on massive beams. These impressions were my prelude to being ushered down a long dark passage for interview by Grose-Hodge, a man whose kindliness towards an over-awed ten-year-old remains an undying memory.

But the Great Hall was not always sombre. Its face changed with the seasons and the terms. It could be festive, as at the Carol Service, when the Chapel choir, in robes, occupied the platform; or colourful, as on Speech Day when academic hoods and aldermanic robes were on show; and even though the Hall faced north, sunlight would shine in through the dormer windows on the south side of the roof, casting surprising pools of light on the north wall; and late on a summer evening sunlight would stream through the main windows and light the east end of the Hall.

The Great Hall was used much more frequently as a place of assembly for the whole School than it is today. Nine o'clock prayers were held there every morning, one o'clock prayers ended morning school on the three games afternoons, and twice a week the Upper School used to spend half of the thirty-minute morning break in the Hall either for hymn practice or for musical appreciation. Occasionally the last period of the morning was marked by a long bell, and this was the signal for an extra assembly. Usually this was because the School was to have an extra half-holiday in honour of some outstanding success, but sometimes it was to receive a general berating from Grose-Hodge.

The seating in the Hall was arranged to face west, and by the time I joined the Upper School all the old bent-wood chairs had been replaced by the oak chairs with rush seats that had been given under the Oak Fund. We were appalling vandals in those days, just as boys are now, and we soon learned that it was easy to administer an unpleasant kick to the boy sitting in front through the pliant rush fibres of his chair seat. There was not much time to do this in Prayers, however, since some five minutes before Grose-Hodge appeared on the platform the Monitors would come into the Hall and line the side and centre aisles. Any form of misbehaviour was swiftly checked by a neatly aimed flick from the end of a long cane.

The Great Hall was essentially the heart of the School and the source of its corporate sense and identity. Indeed the beginning of term assembly used to be attended by all three departments of the School. The Upper School sat on the ground floor, the Lower School sat on the tiered seating in the east end galleries, whilst the members of the Preparatory School stood in the galleries at the west end above the platform and squeezed their bare knees through the wrought-iron railings. But the Hall was used for many other occasions, both formal and informal. The high-point of the year was, of course, Speech Day. The dignitaries would process from the Common Room, the north-

(J. Sylvester)

east ground floor room, along the gravel and enter by the rarely used north entrance. The majority of boys had to sit or stand in the galleries, and after the ceremony, when the procession had left the Great Hall, squadrons of Speech Day programmes carefully folded into paper aeroplanes would float down at various speeds from the galleries.

It was the setting for entertainments. The ugly projection box in the middle gallery came into service perhaps twice a term, and since all cinemas were without exception out of bounds the Great Hall would always be packed with eager boys when a film was to be shown. Grose-Hodge did not seem to approve of such film shows; one felt he accepted them as a necessary concession to boys' lower tastes. Certainly the relevant notice in Prayers on Saturday always referred not to a film but to 'an exhibition of motion pictures'. Concerts and plays meant an enormous upheaval, for the platform, which was only some twelve feet deep, had to be extended out on to the floor of the Hall, with the result that the rows of chairs were horribly cramped together. Alarmingly rickety tiered seating would be erected once a year for the choral concert, whilst in the Christmas term a complete proscenium arch spanned the west end for the Dramatic Society's production, and in the days before it became fashionable to act in front of plain curtains the Art Department and the stage crew provided sets of amazing complexity despite the confined space. Other annual events were the Quadrangular Tournament between Eton, Dulwich, Haileybury and Bedford, when the centre of the Hall was cleared for the boxing ring, and the Assault-at-Arms which was a display of gymnastics. For this climbing ropes had to be fixed to one of the beams above the platform, which meant that the senior physical training instructor, Sgt-Major Jackson, had to climb out by a ladder on to the beam and thread the ropes through metal eyes round it. This was a task that few would willingly have undertaken, given the forty-foot drop that awaited a careless move. But a number of boys over the years risked their necks by walking the beams, and one left a matchbox at the mid-point of one beam to prove his feat.

The major disruption to normal use of the Great Hall each year came in the period when public examinations were sat. The floor of the Hall was cleared of chairs which were stacked in the galleries, and fold-up desks were fetched out of storage and arranged in long lines for the candidates. The period change bell was switched off for the duration of the exams, and a design feature of the building which Phillpotts had insisted on allowed boys to reach classrooms without coming into the Hall and disturbing the concentration of the candidates. Hours of tedious invigilation sessions were the lot of the Staff who patrolled quietly amongst the desks, carefully avoiding those boards on the east end platform that creaked. Whilst the Hall was being used in this way Prayers were held on the south side of the School, but despite all good efforts to make the best of things, this was an unsatisfactory arrangement; hymns were muted, prayers and notices were blown about by the wind and could not be heard clearly. We were always glad when the Hall came back into use again.

But for all the virtues of the building even the most sentimental is forced to admit that it had its limitations. The Great Hall could never be adequately transformed into the modern sort of theatre the Dramatic Society needed, and its use as an examination hall was a source of considerable inconvenience to the School. Classrooms were poorly heated, ventilation was not easily achieved, the floors were of bare timber, some classrooms were much too large, some were divided in two by partitions that were far from soundproof. The building had been designed for a style of teaching that had passed, and as the years went by it became increasingly obvious that the building could not be adapted as it stood to take advantage of the innovations in teaching that modern technology was making possible.

The building was only eighty-seven years old when it was gutted by fire. Many people thought it was much older and this is perhaps a tribute to the solidity and permanence it seemed to symbolise. Its architect and builder, its masters and pupils of the first years had long since gone to their graves, but the classrooms, the echoing corridors and the Great Hall seemed destined to stand for many generations, linking in one continuous chain the pupils and teachers who would pass through it. It was the largest, most sumptuous building to be erected in Bedford in recent times, and the conflagration that raced through it the most dramatic ever witnessed in the town. For those who stared in bewildered disbelief at the smoking ruins all seemed lost. But the fire had not reckoned with the solidity of the masonry that Robins had designed and Spencer had built. The walls still stood. They were scarred, but they were sound.

Chapter II

The Fire, March 1979

Saturday, 3 March 1979, began like any other Saturday. After morning school the Main Building was locked up as usual though access was obtainable, as always, through the west door.

At 2.20 p.m. a senior boy, Christopher Smart, in his role as head projectionist, organised the Great Hall for the film show later that evening and played the film over to see if there were any snags that might interfere with the entertainment. Either he or some of the other boys involved were around for most of the time preceding the showing of the film. From 5.00 to 6.00 p.m. detention was held in A12. Between 6.00 and 7.30 p.m. there was a Lower School play rehearsal of a drama based on the life of Brunel. This was held in C1. Then at 7.00 p.m. came the showing of the Upper School film attended by about a hundred boys under the supervision of Mr John Osman, after which the Great Hall was tidied up ready for Monday morning assembly, fourth forms only. Concurrent with the film was the Modern Languages Society play in the Memorial Hall. The play, *Orphée* by Jean Cocteau, was directed by Mr John Pidoux who locked up when the play was over, shortly before the film ended at about 9.00 p.m.

At 9.30 p.m. the Head Porter, Mr Doug Simms, began the final locking-up procedure, concluding just before 10.00 p.m. The last thing he did was to throw the main electrical switch which turned off all the lights in the building. It did not, however, affect the mains power supply.

Thus by 10.00 p.m. the Main Building was empty and locked up for the night.

Various people, boys and others, came and went in the School Estate during the next hour and a half. The last of these would seem to have been two Burnaby boys, Richard Howe and Hamish Ferguson, returning shortly before 11.30 p.m. from the Dame Alice dance. And the very last person on record as having been in the vicinity of the School was Mrs Frost who works in the School Dining Halls. She drove along Pemberley Avenue at 11.40 p.m. and stopped to drop off a friend.

At this time there was no sign of any fire.

The next significant time is 11.58 p.m. when the Post Office emergency switchboard received the first of thirty-six calls to a fire at Bedford School. This timing is precise, as also is the entry in the log at the Fire Service HQ. There is in fact a discrepancy of three minutes which can be accounted for in many ways not worth considering here. The entry reads as follows:

Original call for incident at Bedford School, Bedford.
(time of call: 0001 hrs. 4 March 1979) (male caller)
(Call originated from telephone kiosk Kimbolton Road, Bedford)
Transcript from tape recording:
G.P.O. Connecting you to Bedford 51467.
Caller: Hello.
F.S.   Fire Service; can I help you?
Caller: I'm in Kimbolton Road. I want to report a fire in, I think, Bedford School Chapel.
F.S.   That's Bedford School Chapel. That's Kimbolton Road?
Caller: That's right, yes, near St Peter's Street, that end of Kimbolton Road.
F.S.   Can I have your exchange and telephone number?
Caller: Bedford 51467.
F.S.   What's on fire, is it actually in the building?
Caller: That's right, yes. It's the roof. Whopping great flames coming out of roof and lots of smoke. I think it's Bedford School Chapel in Bedford School grounds.
F.S.   We will get someone there.

End of call

(From Fire Investigation Report: Appendix 2.)

This was the first of the thirty-six calls logged by the Post Office. A large number of people who telephoned to raise the alarm found the lines jammed.

If the development of events around midnight seems confused, that is because the fire grew and spread with exceptional and indeed extraordinary rapidity. After the event one attempts to fix the precise time at which something happens, but the timing is more often than not approximate. What happened seems to have been as follows.

At about 11.55 p.m. the two Burnaby boys previously mentioned, Howe and Ferguson, burst into the bedroom where Mr and Mrs Davidson were now in bed. The blaze was already dramatic and so immediate action was taken in alerting all the boarders in the house.

At about the same time Bill Amberg, Head of Kirkman's, invaded the bedroom of Mr and Mrs Rawlinson with the same news.

At about the same time Mr and Mrs Sylvester, as far away as Farrar's, also in bed, actually heard the fire. As it happened there had been a fire at

(Bedford County Press)

Gammans, a shop in the High Street, earlier in the week and Mr Sylvester had been struck by the noise it made and in particular by the sound of spitting, cracking, falling tiles. These were the sounds that roused him now.

Professor and Mrs Constable at 20 Kimbolton Road were the second callers to alert the Fire Brigade. They discovered the fire when Mrs Constable's brother who was staying with them went out to his car to see that his dog was all right. He returned to the house with the laconic remark, 'You'd better come out if you want to see Bedford School on fire.' The time must have been around midnight.

To return to firmer timings, the Fire Brigade first received news of the fire at one minute after midnight (see transcript above) at their Headquarters in Kempston. They contacted the Barker's Lane Fire Station where Sub Officer Wheeler, receiving the message, immediately set off with two fire-fighting vehicles in his charge towards Bedford School. His route went via the Embankment, from where he was astonished to see, not the expected glow of a fire in the middle distance, but actual flames. Professor Constable says that when he looked out, probably about this time, the flames must have been thirty to forty feet higher than the roof, itself in the region of ninety feet above the ground. One must therefore visualise flames leaping up to one hundred and thirty feet in the air, visible in may areas of Bedfordshire. Indeed, Sub Officer Wheeler records that a fire-fighting vehicle coming a while later, from Luton, could see the fire clearly from the hill above Barton-le-Clay. He himself, proceeding along the Embankment, not only saw the flames but heard the roar of the fire. It was obviously a situation of major emergency.

Imagine, then, his frustration on arriving at Glebe Road to find the iron gates to the School padlocked. Included in his equipment was a pair of bolt croppers, but the time taken to shear through the steel chain seemed an age. The fire tender carried only four hundred gallons of water plus a ladder. He saw at once that the situation was desperate and so he radioed the base at Kempston to state 'Main School Building of two floors well alight', and to request immediate assistance. He then directed two water jets at the north front, but with the height and grip of the flames on the structure this was pitifully inadequate, so he confined himself to a single hose in order to get more pressure and hoist a higher jet. When within minutes the second fire tender arrived he ordered it to bear on the south front of the building. However it was clear from the first that with a moderate, though wet, westerly airstream fanning the flames, the School Armoury with its store of CCF ammunition and explosives, and the Science Laboratory with unimaginable store of combustible materials were in danger: and beyond these the houses in Glebe Road.

So at thirteen minutes past midnight, a mere seven minutes after his arrival, Sub Officer Wheeler sent a radio request for the police to evacuate the houses in Glebe Road adjacent to the school – obviously a shrewd assessment of the situation.

From this time on fire-fighting reinforcements continued to arrive until eventually every single fire station in the county, thirteen in all, had made its contribution. Surrounding the School there were eighteen fire-fighting appliances of one sort or another including a turntable ladder 30 metres high, one 25.5 metre hydraulic, and two 15 metre platforms. At 00.23 a.m. Chief Fire Officer Haley himself arrived to take charge of the proceedings. It has been said that this is the biggest as well as the most expensive fire ever to have occurred in Bedfordshire.

A point of interest here is that for some time the water supply proved quite inadequate to even begin to master the fire and, as the fire officers themselves have said, the fire was at its height inextinguishable. A lake dumped on the Main Building would at once have been turned to steam and evaporated into the atmosphere. The 400 gallons in the fire tenders were a mere thimbleful. Hoses attached to the mains were totally unable to achieve the required height since the water mains at the hydrant points lacked the necessary pressure. At 00.24 a.m. the Water Authority was requested to increase the pressure. But that took time. In the interim the Fire Brigade's thirsty pumps broached the open-air Swimming Pool but that supply was soon exhausted. Eventually, in the search for suitable pressure, a hose was run all the way to the hydrant by the Granada Cinema.

Since the main centre of the conflagration was beyond control, the efforts of the Fire Brigade were concentrated on preventing the fire from spreading. The east end of the School, the Science Laboratory, and the Armoury were liberally doused in order to keep the exteriors cool, wet, and fire-resistant. This was the priority. Second to this came the application of a similar tactic to the School's west end. The overall strategy was assisted within the building itself by the fire doors at all three levels at either end. Made of oak with steel reinforced glass panels, they miraculously resisted the furnace. The photographs taken afterwards give a clear indication of the devastation on the Great Hall side of these doors, as contrasted with the total immunity from damage on the other side. However, this hidden resistance to the flames could not be known by those outside, fighting to contain the holocaust.

All those who were present bear testimony to the courage and selfless daring of the firemen throughout the operation. The danger derived not only from the flames themselves but from explosion, burning brands and other debris flying through the air, the toxic effects of smoke, and the devitalising effects of heat radiated from the red-hot bricks. Chief Fire Officer Haley's comment on this fire harked back to the Second World War. 'It was just like the Blitz,' he said. The element of danger was very considerable. Mr Sylvester reports that he saw a fireman directing water from a platform high up among the flames by the roof. Suddenly there was a spurt of flame which seemed to engulf him, followed immediately by a pall of smoke which took an age to clear. It seemed impossible that he should survive, but he did, and continued to operate his hose. Similar situations were common. Firemen were active everywhere and it was miraculous under the circumstances that no one was injured: a testimony no doubt to discipline and rigorous training, intelligent organisation and a wealth of past experience. The official fire investigation report comments with supreme understatement, 'Fire fighting operations progressed smoothly'. They did indeed. The fire was contained, it declined in due course, and by the morning all that remained was a scarred, blackened, smoking shell of a building with here and there deep-seated pockets of fire which required the presence of fire-fighting crews until 3.00 a.m. on the morning of Monday, 5 March.

However, at or around midnight on Saturday, 3 March, it was by no means certain that the fire was going to be contained. As has already been indicated, the scale of the fire was considerable. From the moment of the first sighting, and no one is precisely sure when that was, the fire developed with unusual speed and the violence of the conflagration was exceptional.

Looking back, there are reasons why this should have been so. The main structure of the building may have been brick but it was faced in the Great Hall itself with oak panelling at ground-floor level and there were oak passages at either end. The roof was constructed from huge pine crossbeams on which over the years dust had settled. At the time of the fire, according to Mr Doug Simms, the dust was probably at least an inch thick and a perfect conductor for flames to race over the beams as if they were sprinkled with gunpowder, setting alight anything combustible on the way. Moreover, the woodwork had been stained or polished until it was a rich, dark, reddish colour. Assistant Divisional Officer Holloway who conducted the Fire Investigation Report, has described this as 'button polish' and explains that

The fire in its early stages devoured the roof and from the very beginning signalled its presence violently and dramatically. *(Bedford County Press)*

it is not unlike beeswax and, once applied, penetrates into the fibres of the wood. In addition the Great Hall windows were curtained with velvet from sill level right up to the apex of the window arches. Once ignited, these curtains would conduct the flames to the roof where there was a wealth of timber panelling and supports in addition to the great crossbeams. And, finally, all timber in a roof tends to be exceptionally dry and combustible. These are some of the reasons why the fire in its initial stages concentrated in the roof, and from the very beginning signalled its presence so violently and dramatically.

The instance of Sub Officer Wheeler who arrived at the School at 00.06 a.m. has already been cited. He saw flames from as far away as the Embankment. Mr Lazenby of 24 Kimbolton Road, returning home from his daughter's 18th birthday celebration dinner, saw a glow from Goldington Green and, as he turned into Kimbolton Road, seeing his house a black silhouette against a background of flame, thought at first that his home was on fire. Mrs Sylvia Nicoll, living then at 11 Glebe Road, right on the spot, describes the flames as 'leaping up through the centre of the building'. And, of course, well above the roof line, was the fifty-foot spire of pine construction with a thin cladding of copper. The sight of that blazing above the incandescent roof was a sight indeed. It might well be considered as the major spectacle, dominating all the surrounding inferno.

And amazing sights and sounds there were. Mrs Nicoll mentions the horrifying speed with which the flames devoured the roof, the crackling noises, the smells of burning, the showers of sparks fountaining into the air as if in a macabre firework display. Mrs Davidson, watching with her family and boarders from the adjacent flat roof of Burnaby, talks about the loud reports of tiles that split and flew whirring through the air. Mr Lazenby, for his part, talks about the avalanches of tiles released from their hold, slipping down the slopes of what was once a roof. All of those present at this time make mention of the percussive noises made by the tiles.

The moderate south westerly wind created a furnace effect. The roar of the flames could be heard half a mile away. *(Bedford County Press)*

But presiding over all was the flaming spire, fanned into incandescence by the wind, powerful at that height, until, inevitably, the moment came when the wooden supports were eaten through and unable to hold the main structure. The tower tilted and sank into the anonymity of flame that had been the Great Hall. One observer in Glebe Road, Mrs Cartwright, describes its sudden disappearance as 'appalling'. Mr A. C. Fitt (OB), Head of Biology at the School, says that on the collapse of the spire a great fountain of flame shot well over a hundred feet into the air and sparks showered all around. Mrs Struthers of Pemberley Avenue mentions 'a great fountain of sparks and flame'. It was spectacular.

Reactions at this moment were many. The boys on Burnaby roof to whom the stubborn resistance of the spire to flame seemed incredible, raised a thin cheer as it subsided. Others, watching from the School field, were silent or gave an involuntary sigh. At this point it is necessary to state that those watching the fire were not all members of the School or occupants of houses in the immediate neighbourhood. Telephone calls had been made and many people who had no connections at all in the area came to watch, having seen the blaze from afar. There was a sizeable crowd. And among those present there were some who viewed the fire with relish. These included the arsonist who started it and some of his associates. He afterwards admitted that he had been present, and that he had cheered and jeered when the tower fell.

Mrs Nicoll, evacuated from her home which was close to the Armoury, but still watching from Glebe Road, writes, 'I saw two youths in leather jackets watching and almost laughing; they were just passers by. One, laughing, lay down on the pavement and lit his cigarette from a burning ember.'

Feelings, however, ran high. And as the Head Master puts it in his Foreword, bystanders who indulged in ribaldries were soon 'put in perspective'.

It is interesting to review briefly the reactions of others watching. The younger boys, boarders in Farrar's, were given the opportunity by Mr and Mrs Sylvester to see the sights but almost all, after a brief glimpse, found the imperative of sleep too much for them. Miss Prosser, of 14 Park Avenue, talks about 'one of the most traumatic moments of my long life'. For her, 'the feeling of hopelessness and indignation was lifted temporarily when a group of boys raised a cheer as the noble tower sank down to rest among the ashes, and I was particularly moved to watch the destruction of the Great Hall with its chairs which had supported hundreds of people, including me, who had come to enjoy entertainment offered by the School. One cannot help feeling glad that the memory of the old hall will linger on for many years to come.' She had felt 'curiously and involuntarily drawn to join others in sharing the agony of helplessness in seeing something precious destroyed before our eyes.'

Mrs Enid Cartwright, of 2 Glebe Road, had been woken shortly after midnight by the telephone. A neighbour escorted her away from what was thought to be a danger area. 'What a dreadful sight caught my eyes,' she writes. 'It was ghastly to see the massive flames mounting into the heavens with sounds of crashing timbers. It was heart rending.'

Mrs Nicoll talks of 'immense shock . . . one just stood about in dumb horror . . . the worst part seemed to be when the copper tower melted and disappeared as if it were melted butter'.

Many of the spectators had pulled on warm clothes over pyjamas and were, like Mr and Mrs Struthers, 'eccentrically dressed' and talking in 'hushed, shocked tones'. She continues: 'We gazed, fascinated, at the frightening sight . . .. We have no direct connection with the School, but perhaps felt the worse for that – almost as though we had witnessed the immolation of a kindly and benevolent neighbour whom we were powerless to help.' Mr Lazenby concurs: 'There was that awful sickening cheer when the spire collapsed . . .. The sense of shock, horror, and disbelief was quite remarkable.' Ms Juliet Adams, née Boddington, who as a girl lived next to Pemberley but at the time of the fire was in Bushmead Avenue, watched from the 'far end of the Lower School pitch. The only other people there seemed to be from Redburn. I saw and heard the copper tower fall; the crash epitomised the whole episode.'

The final disintegration of the copper spike. *(Steve Beesley)*

But with the fall of the spire the night of flame was by no means over. Indeed, although to Mrs Davidson watching with the Burnaby boys the spire seemed to hold out for what she thought must have been two hours, Sub Officer Wheeler thinks that it must have foundered within forty minutes of his arrival with the first fire-fighting tender. Timing is often highly subjective. Some time, however, after its spectacular collapse, Mr Fitt accompanying his son Peter and some friends round the School witnessed an exceptional sight. All at once, together, the windows on the south front exploded, glass showering everywhere and, accompanying this explosion, tongues of flame thirty feet long leaped out of every window. This brief emanation of flame was succeeded by a secondary phase in which the whole south front of the School showed black with the yellow squares of the burning window frames dissonant against the deeper red of the main fire.

Eventually, however, it became clear that the operations of the Fire Brigade were succeeding in their limited objective of containing the fire. The Fire Officers themselves concede that there was nothing they could do to put out the central core of flame. At some stage, then, during the early hours when it was clear that the fire at the west end had been contained, the Head Master received permission to enter the building in order to salvage what he

could. His priority was to secure the documents relating to the boys. Each boy has a file which records all the data relevant to his career from the first letter of application by his parents, through to the details of his examination results and/or examination and career prospects, together with medical records etc. It was desirable, therefore, if possible, to carry these to safety since with an Upper School of 720 boys the labour of rebuilding the files would be formidable.

With a small party of assistants the Head Master succeeded in this task. Looking back, he comments that it was remarkable no one was injured. Although the roofline must have been practically denuded there was still the occasional hum, whirr and smash as tiles from the west and east ends of the building spun from what was left of the roof. Inside the Head Master's study, Mr Fitt records, 'the window frames, curtains and everything else were burning pretty vigorously'. Nevertheless the rescue party led by the Head Master succeeded in salvaging almost all the necessary documents. Apparently, although the doors into the Great Hall were, obviously, gone, there was no undue feeling of heat inside the building. Possibly this was because the heat radiating from the bricks outside was so intense that no increase was perceptible inside. The Head Master's study, the secretaries' office and the Bell Room were, of course, full of smoke and very wet, but all the documents in the steel filing cabinets were protected and survived. Almost certainly the preservation of these papers was due to the effectiveness of the fire doors. The only documents lost were those on desks awaiting typing and filing and those stored in wooden filing cabinets.

Once the documents were in safety, the rescuers thought of the valuable offset litho machine, a new acquisition, lodged half-way up the west staircase between the ground floor and the Lower School Common Room, in what used to be called the 'Stationery Room'. This heavy item was manhandled half-way down the stairs when Mr Fitt who was, so to speak, the downward man, missed his step and let go of his end. He may have been soaked already but was now further subjected to an inundation of liquid from the tilted machine. The offset litho was severely damaged by the jolt and a day or so later Mr Fitt's socks and trousers, thanks to the offset liquid, developed a spectacular series of holes. The irony of this incident was that, had the machine been left where it was, it would have received no injury: and Mr Fitt would not have lost his socks and trousers.

It was at this time, after 3.00 a.m., that the Chief Fire Officer issued his statement that fire-fighting operations were progressing smoothly. The message continued: 'Stop for Bedford School, Bedford, a building of 3 floors approximately 75 metres by 25 metres used as a day school premises. All 3 floors and roof involved in fire. 90% of the building and contents severely damaged by fire.' From then on, but slowly, the fire began to decline. As it did so it relinquished some of its spell on the watchers, who began to drift away.

By about 4.00 a.m. the spectacular main act of the drama was over: the fire continued to burn, though with ever decreasing vigour. Those who had come to watch had departed. By 5.00 a.m. those connected with the School had also retired, leaving the firemen to their unenviable task.

But sleep, for many, was no more than a brief period of restless blankness sufficient to restore a fund of energy for what was to ensue. As the Chief Fire Officer said, this fire was like the Blitz: not only in the fierceness of its incendiary period but also in the emergency it had created.

Among those who had watched the fire, boys in particular, there had been some speculation about the next few days, weeks, months. After all it was abundantly clear that the main centre of teaching was destroyed. Whatever course of action might be adopted, it was inconceivable that the complex machinery of School life would continue. The likelihood was that priority would be given to providing facilities for the 'A' and 'O' Level candidates to continue their studies; and the rest of the boys sent home. This solution, however, never crossed the Head Master's mind. His thoughts concentrated on finding locations for 30 classrooms.

Ian Jones has said that at this time, for some unknown reason, he had a compulsion to walk round the burning School: time and again he paced round the burning remains, not consciously applying his mind to specific detail though perhaps general outlines were beginning to formulate themselves in detail in his thoughts. The destruction of the Main Building meant the destruction of 30 teaching rooms. Where could 30 teaching rooms be found?

He decided that he must convene a meeting at the earliest possible time in order to discuss immediate action. The time fixed was 10.00 a.m.; the place, School House, those attending were to be:

H. P. Shallard, Chairman of the Bedford School Committee of the Governing Body;
R. N. Hutchins, Clerk to the Harpur Trust;
D. R. McKeown, Development Manager of the Harpur Trust;
D. R. Mantell, Bedford School Estate Bursar;
H. T. Inskip, Bedford School Surveyor;
Mrs Anne Rhodes, Head Master's Secretary;
Mrs Jennifer Jones, Head Master's wife;
M. E. Barlen, Vice Master;
P. R. O. Wood, Usher;
J. H. Davidson, Housemaster of Burnaby, Senior Housemaster; and
F. M. Fletcher, School Registrar.

Of the above, the people on the spot were notified immediately during the fire.

The Head Master attended Communion as usual in Chapel at 8.15 a.m. before visiting the Dining Halls where those boarders who had not gone home on the 'exeat' weekend were having breakfast. Then he returned to

School House for his own breakfast and to prepare for the meeting at 10.00 a.m. The results of that meeting were as follows.

1. Teaching areas were reorganised as follows:

A6	Mr M. E. Barlen	became	Pemberley 1
A7	Mr H. D. Galbraith	"	2 Burnaby Road
A8	Mr M. J. Rawlinson	"	Kirkman's
A9	Mr D. C. Bach	"	Physics Project Lab. (S7)
A10	Mr P. G. Braggins	"	Pemberley 2
A12	Mr J. H. Davidson	"	Burnaby
A14	Mr P. J. Coggins	"	F8
F8	Rev. M. D. A. Hepworth	"	Talbot's 1
B1	Mr T. J. Machin	"	Fitzpatrick
B2	Mr P. G. Bossom	"	Talbot's 2
B4–B7	Lower	"	Inky 3: Nash's 2
C1–C7	School	"	Craig 2: Art 1
B8	Mr G. F. W. Pleuger	"	Downstairs Drawing Office
B9	Mr P. S. Wiser	"	Howard's
B10	Mr R. J. Robinson	"	Temporary Classroom 2
B11	Mr J. B. D. Osborne	"	Kirkman's 2
B12	Mr M. P. Stambach	"	Farrar's 1
B14	Mr T. J. Elliott	"	Rifle Range
B15	Dr J. F. Foulkes	"	Farrar's 2
B16	Mr F. M. Fletcher	"	Talbot's 3
C8	Mr D. H. Bullock	"	Dining Hall 5
C9	Mr D. W. Jarrett	"	Pavilion
C12	Mr T. A. Riseborough	"	Biology Projects Lab.
C13	Mr J. F. W. Pidoux	"	Art Block
C14	Mr A. E. Barlow	"	Music School Hall
C16	Mr D. P. C. Stileman	"	Dining Hall 6

(Additional note: study periods should be taken in Boarding Houses or the Chapel; reading periods in the Chapel.)

2. A letter sent by the Head Master to all parents is printed opposite.

3. The notice opposite was placed at the various entrances to the School Estate so that boys and masters in ignorance of the fire might know where to go, and when.

4. It was established that the Head Master's study, the secretaries' office, the Bell Room and reception room for parents and visitors should be located in 6 Burnaby Road; and that the Upper and Lower School Staff Common Room should be housed in the Music School if the Director of Music, Mr Amos, agreed.

5. In the first instance the offices of the Harpur Trust would, so far as possible, meet immediate needs, providing stationery, typing and

(Steve Beesley)

Founded 1566

THE BEDFORD CHARITY
(THE HARPUR TRUST)

The Harpur Trust Office, 101 Harpur Centre, Bedford, MK40 1PJ

Telephone: *Bedford (0234) 42424*

IN REPLY
PLEASE QUOTE: ....

Your Ref: ....

Clerk of the Trust   R. N. Hutchins,   LL.B., D.P.A., Solicitor
Finance Officer   M. J. Ferdinando, I.P.F.A.
Development Manager   D. R. McKeown   F.R.I.C.S., F.I.Arb., F.F.B.

4th March 1979

FROM THE HEADMASTER OF BEDFORD SCHOOL

Dear Parent,

You will know that a fire completely gutted the Great Hall and Main Building of the school on Saturday night. Fortunately no one was injured and thanks to the Fire Service the fire was eventually contained and did not spread to other buildings.

You will, I know, appreciate the considerable problems we now have, but it is our firm intention that school will continue as normal with the minimum disruption.

Our short term priorities will be to provide temporary accommodation for the classrooms we have lost, and to assess the cost of the contents, both personal and school, destroyed.

For the remainder of this term the classes taught in the Great Hall building have been relocated within the school estate. The school office will operate from 6, Burnaby Road. The Upper School Common Room has moved to the Music School, and the Lower School Common Room will join the Preparatory School Common Room.

We have already received a large number of letters expressing sympathy and offering help and I shall never be able to thank people enough for this either on behalf of the school or personally. This support is a very great strength to all of us.

Internally the school has a very fine staff and a first rate set of boys, and all of us will now be working harder than we have ever worked before to get the school back into first rate order in as short a time as possible.

I will, of course, let you have further information as this becomes available.

Yours sincerely,

C.I.M. Jones
Headmaster

BEDFORD SCHOOL

NORMAL SCHOOL ON MONDAY MARCH 5TH AT 9.00 A.M.
PREPARATORY SCHOOL ASSEMBLE AS NORMAL IN THE INKY
LOWER SCHOOL TO THE GYMNASIUM, PLUS ALL LOWER SCHOOL MASTERS
IVTH FORMS TO SCIENCE BLOCK S.18 PLUS ALL FOURTH FORM MASTERS
REMOVES, FIFTHS, SIXTHS TO THE CHAPEL PLUS ALL OTHER UPPER SCHOOL MASTERS
PARENTS EVENINGS LOWER SCHOOL MONDAY MARCH 5TH IN DINING HALL 6
PARENTS EVENINGS FOR R1, V2, V3, V4, V5, V6, V7 DATES AS ARRANGED BUT IN DINING HALLS 5 & 6

duplicating facilities and so on. It was also necessary to communicate at the first possible opportunity with the Oxford and Cambridge Examinations Board in Cambridge in order to safeguard the examination prospects of the 'A' and 'O' Level candidates who had lost notes which would have been used for revision.

6. The Bursar had already arranged a meeting for 11.30 a.m. at which, obviously, insurance matters affecting the loss of personal property as well as the larger issue of the School itself would be discussed and action determined.

7. It was decided that the Bedford School Committee of the Harpur Trust should meet on Thursday, 8 March, to consider in detail the possible courses of action.

Later it was possible to assess accurately the extent of the disaster. The Great Hall together with all the form rooms on its south side was totally destroyed at every level. All that remained were the galleries of concrete superimposed on a cast-iron grid which was of course indestructible by fire: all the wooden floors of the southern classrooms had burned through and had collapsed. The whole of the roof had gone but the brick chimneys and wall ends remained upright, looking far from safe. All the rooms and lobbies in the east and west wings, however, remained more or less intact, though obviously affected by smoke and the deluge from the firemen's hoses. Effectively the building was gutted.

Throughout the daylight hours citizens of Bedford, hearing the news, came to gaze and, many of them, to mourn. Comments expressed in letters include: 'I stood there silently, completely overwhelmed'; 'so many memories gone – the Head Masters' portraits, the draughty hall with memories of Christmas plays, BBC concerts and Hobbies Exhibitions'; 'I visited the ruins with a friend and stood, like others, dejected, uncomprehending, wisps of dust and smoke eddied in the light of that otherwise pleasant early Spring morning'; 'in the morning it was a dreadful sight to see the lovely school just a disastrous ruin and the grounds and Glebe Road a mass of water and debris'.

And now a word about the weather. During the night of the fire there had been a moderate westerly wind and it had rained continuously. With dawn the rain ceased and the weather became mild and sunny, facilitating the reorganisation of equipment. Boys were able to carry desks and chairs and tables about without undue discomfort. On Sunday night the rain set in again. The dry spell during the day had been providential.

Inevitably, given the scale of the conflagration and the westerly wind, the whole area to the west of the School was littered with charred fragments of wood and burnt pages from books. There were also a number of fire engines and a presence of fire-fighters and police. The area immediately round the building was kept clear since there was a very real danger of falling debris. No one at this stage knew how firm the chimneys and other skyward projecting contours of masonry might be. It was not until some days later

The Great Hall: the morning after the fire: total desolation. *(Bedford County Press)*

The view of the Great Hall from the gallery. *(Bedford County Press)*

that the structural engineers made their surveys and pronounced the building safe. Indeed so thoroughly had the Victorian craftsmen done their work that when he first met the Staff Planning Committee Sir Philip Dowson, with mild overstatement, described the building as 'atom-bomb proof'.

Now that the roof had gone, and therefore the proportions of the building significantly altered, one realised what an enormous size it was. This was particularly striking when walking along the south front. 'What a barrack', someone said. One realised then, forcibly, what should have been appreciated before, that the Victorian-Gothic protuberances – tower, spire, eye-brow dormer windows – were not merely profusions of Victorian confidence and optimism but had an important aesthetic function. Without them the structure lost its overall harmony and became 'a barrack'. A comparison between the present roofline, viewed from the north, with photographs of the original roof emphasises this point. Without the seven dormer windows to break the line, the great span of the present roof is emphasised.

On Sunday afternoon Mr Wood organised the masters and boys who had volunteered to help with the disposition, in the new teaching sites, of the various items of School equipment that were needed – desks, chairs, blackboards and so on, which were stored on the School Estate and used only for examinations. An interesting document surviving this time, in Mr Wood's handwriting, tells of furniture that might be salvaged from within the Main Building: e.g. from B16 (Mr Fletcher's room), '12 tables: 20 chairs: 27 lockers (a bit hot): filing cabinet?' A firm 'No' is pencilled against the locker entry!

This Sunday was an 'exeat' Sunday and so boys who for various reasons had stayed at School were, naturally, phoning their parents with the amazing news. Mrs Rawlinson, writing about Kirkman's boys, says: 'The general trend of these calls was, "Mum, you're not going to believe this but the school burned down last night!" . . . pause for Mum's exclamations of disbelief, followed by filial reassurances that all was well. In the evening, those returning from exeat could not believe their ears when they were greated with the news. More phone calls on the same lines. The story made the ITN news, with pictures, later on Sunday, and the national papers on Monday. Further frantic phone activity, this time incoming calls from our own family and friends as well as those of the boys: it hadn't been clear to anyone not familiar with the layout of the School that some areas were untouched.'

Mrs Davidson, on the same theme, writes that 'an Old Boy returning his son to Burnaby was confronted by the still smouldering shell of the School building. He turned away with tears streaming down his face, and his son followed suit. To cheer the boy I said, "If you had been here with all of us you would have found it tremendously exciting." I realised at once from his father's face that I had said the wrong thing.'

On the Monday morning, boarders who had been present while the School burned down as well as those who had not, day boys who had heard the news as well as those who had not, even some of the teaching staff who had been ignorant of the catastrophe as well as those who had been in the thick of it, assembled in accordance with the notices posted at all entrances to the School Estate.

In the Chapel where the majority of the Upper School were gathered there was silence as the Head Master made his way to the pulpit. He spoke as follows:

Stunned, shattered, horrified, shocked, bewildered, tragic – these are just some of the words which can only inadequately describe our present feelings.

All of us will be reminded of the events of the past weekend for a very long time to come. They will leave deep scars on the hearts of a vast number of Bedfordians, and ours particularly as long as the shell of the building is there to remind us of one of the finest school buildings in this country.

We must remember, though, that life is also about *people*, and fortunately this School has a very fine staff and a first rate set of boys, and *together* we must all set about the task of dealing with the problems that have been created for us.

Our first priorities have been to ensure School continues.

The second will be to attempt to assess the value of the contents belonging to both Staff and boys and School, and details of this will follow later this week.

Then we shall need to make our permanent plans for the future.

What can you do to help?

First, react with determination and courage to the challenge that faces you individually and as a community.

Second, remain optimistic and cheerful and helpful.

It will mean all of us working harder than we have ever worked before and we cannot escape that fact. The eyes of a large number of people both nationally and in Bedfordshire are on us, and we can take great encouragement and great strength from the vast number of people who have already offered their help.

We want the next few years to be the best this School has ever had, and when I say best I don't mean best in terms of Oxbridge, 'O' and 'A' results, Henley, Rosslyn Park, Bisley and so on – that's the icing on the cake, and it's very good indeed, but it is what is underneath that counts. So I mean best in terms of what each one of us can achieve – at 'O' and 'A' level, and here let me reassure all 'O' and 'A' level candidates that we will be preparing special circumstance forms for all candidates explaining to the Examining Boards exactly what handicaps you have been under. It means best too in everything we do in games, in music, in drama, in general standards, in all our activities and interests.

We want people to say of masters and boys and everyone at Bedford that they have *big hearts*.

So aim for the *sky* in your work, in your games, your music, your drama, and all your other School activities.

Hold your heads up high both inside and outside the School, during term and in the holidays, so that when we look back on this darkest of hours we can be justifiably proud of our efforts and the cooperation, which, providing there are no weaknesses in the chain, will pull us through.

GOOD LUCK TO ALL OF YOU!

After this came the practical details. The Head Master listed the new locations where teaching would continue until the end of term. Assemblies would be held every day in Chapel except for Wednesdays and Fridays. The Head Master went on to confirm that Parents' Evenings would be held as arranged but in the Dining Halls instead of in the Great Hall. He confirmed also that the whole pattern of School life would continue unchanged except for the new dispositions necessitated by the fire.

There was no hymn. The meeting ended with a prayer and the School dispersed with conflicting thoughts and feelings. Certainly there was excitement and the stimulus of a fresh and unexpected challenge. There was also foreboding at the prospect of the demands ahead. Fortunately the end of term was only two weeks away.

For those who know about school it is not difficult to imagine the welter of detail which had to be sorted. Priorities had to be established and connections improvised for the severed strands which should have linked Saturday, 3 March, the end of one week, with Monday, 5 March, the beginning of the penultimate week of term. The end of term is a time of heightened activity for all except, possibly, the boys themselves. It was therefore vital to have an immediate meeting of the whole Staff to ensure that all the essential services could be continued; and also, following that, a meeting of Heads of Departments to review the way individual subjects had been affected by the fire and to take whatever measures might prove necessary. These meetings were held within the next three days; the problems were faced in detail, and in various ways resolved.

End of term for boys and parents alike means reports, those sometimes unwelcome assessments of application, achievement and conduct during the term. Who, after all, is not aware of his own occasional lapses from virtue?

The panelling shows how effective the fire-doors were. *(Bedford County Press)*

The Head Master with Mr A. C. W. Abrahams, Chairman of the Harpur Trust. *(Bedford County Press)*

Or not apprehensive that he may not receive a proper recognition of his efforts? However, it is right that parents should be informed of the performance of their sons and this was one of those severed strands of communication that had somehow to be made good, since most report books had been destroyed in the fire. Mr Douglas Butler, the master in charge of the School Printing Club, and Simon Frankish, its President, devised, and with the unremitting assistance of the rest of the club printed, substitute report cards which made good the loss. More than that, to replace further losses, the Printing Club had, by the Friday following the fire, supplied the following impressive items of stationery:

> 300 VIth Form Report Cards
> 300 R & IVths Report Cards
> 500 Head Master letterheads
> 500 Registrar letterheads
> 500 Common Room letterheads
> 1,000 Notice for Prayers slips
> 1,000 Health Certificates
> 50 Absentee Forms
> 80 'Keep Out – Danger' signs
> 500 Registrar compliments slips
> 1 CSU order pad
> 300 OB Dinner tickets (venue changed)

If the above is one example of the ways in which people responded to the crisis, the following account is another.

The fire had aroused widespread interest locally, and, from the Sunday onwards, many people in no way connected with the School came to look and wonder at the wreck. This was quite understandable and the expressions of sympathy and offers of help were heartwarming. However, the security aspect had to be considered. With no doors remaining there was no way of preventing curious people, souvenir hunters maybe, from entering the building after dark; possibly causing further damage (though that might have proved difficult), or injuring themselves (more likely). For this reason a security patrol of Kirkman's boys with Mr J. B. D. Osborne in charge and under Bill Amberg (Head of House) patrolled on the school field from 9.00 p.m. to 1.00 a.m. each night for the remainder of the term. It is interesting to look back at some of the instructions given.

> Points to bear in mind:
> 1. No weapons.
> 2. No pursuit.
> 3. No guests.
> 4. No corps uniform.
> 5. Concentrate on Main Building, not School Estate.

Mr R. N. Hutchins, Clerk to the Harpur Trust, with Sir John Howard. *(Bedford County Press)*

At this time a tremendous number of people offered help. Canon David Fricker provided chairs from St Peter's Church Hall; the High School, St Andrew's School and the Prison provided much needed school furniture; the Eastern Electricity Board provided electric heaters; Mr Anthony Ormerod gave a duplicating machine for Common Room use; Mr Brian Howard helped to furnish 6 Burnaby Road as a school office; and, through the good offices of Mrs Rawet of the School medical centre, her husband provided the office with a telephone connection in the middle of Burnaby Road. Within a few days Mrs Pemberton, Headmistress of St Andrew's School, conveniently adjacent, made room for two classes so that boys who in the first days after the fire had attended classes in the Dining Halls, could now make a hopeful pilgrimage up Pemberley Lane.

Thus the teaching programme could continue. The games programme was, of course, uninterrupted and the term was successfully concluded without the loss of any activities.

It is now necessary to attempt to cover the most unpalatable aspect of the fire – how it came to be started.

In charge of the police investigation was Superintendent (then Detective Inspector) Roy Rogers. He has said that the major problem in trying to discover the cause of a fire is that, by its very nature, any evidence there may be is likely to be destroyed. However, the make-up of the human mind is such that, should the cause be arson, the person who lit the match feels a compulsion to be present and to watch his handiwork. For this reason Superintendent Rogers likes to be present at a fire in case he can pick up clues from the behaviour of one or more of the people watching it. On this occasion he was certainly aware of a discordant element amongst the large number of spectators, but was nevertheless unable to feel positive that this fire was due to arson. In spite of that, on the Sunday morning he called in Dr Michael Fereday, a forensic scientist from the Home Office Forensic Science Laboratory at Aldermaston, to search the remains of the building and make a report of his findings. These were negative: there was no evidence of arson. Superintendent Rogers therefore briefed the Press that arson was not suspected.

In his deepest mind, though, the Superintendent may not have been totally convinced by the lack of evidence. For one thing, for a period of two months starting shortly after Christmas there had been an unexplained increase in the number of fires in Bedford, and these fires seemed to be concentrated in an area north of the river. Moreover, he worked in close collaboration with Assistant Divisional Officer Holloway of the Fire Service in whom the conviction was growing that there was an arsonist at work. Assistant D/O Holloway personally undertook a searching and comprehensive investigation into every aspect of the Bedford School fire, interviewing, with police assistance in some cases, every available person known to have had any connection with the building in the hours preceding the fire. But these investigations yielded nothing beyond the obvious possibilities, namely: failure properly to extinguish smoking materials; electrical faults; malicious ignition by person or persons unknown. And of these possibilities, no single one had a stronger claim than the others.

In the event, the true cause of the fire was discovered by chance. A member of the Fire Service overheard a conversation in a pub which made it quite clear that the speaker was boasting of his exploit in setting Bedford School on fire. The fireman reported to Asst D/O Holloway who in turn notified Superintendent Rogers.

Thus the name of Ian Ludman came to the ears of the police who, of course, knew of his previous criminal record. But at this time there was no actual evidence to connect him with the fire.

The precise date of this information is not altogether clear, but it was, significantly, after the next incendiary incident, the setting on fire of the Chicken Bar-B-Q in Harpur Street. Also important, it transpired, was the fact that the premises had just been burgled and a quantity of frozen foods removed before the fire was started.

Superintendent Rogers therefore instigated and personally took part in a search of the back areas, dustbins included, of the addresses which he knew were frequented by Ludman, in particular that of his parents in Goldington and that of Flat 6, 8 St Andrew's Road, where Ludman was known recently to have been. In the very early hours of 22 March, in the dustbins of Flat 6, he found the evidence he needed: the wrappers of catering packs of frozen foods normally purchased only by retailers. On the basis of this evidence he obtained a magistrate's warrant and at 8.40 a.m. entered the flat, roused the occupants and conducted his search. He had not far to look. The refrigerator was crammed full of catering packs of frozen food, and a refuse bin held the cardboard remains of more frozen food. Superintendent Rogers therefore arrested the three occupants. Ludman was not present but the evidence suggested that he had been staying there, and newspaper cuttings referring to the fire at Bedford School were discovered, as well as other property later discovered to have been stolen from the School.

At first these men were unwilling to give any useful information. They could not, however, deny the connection between the frozen food in their possession and the fire at the Chicken Bar-B-Q, nor their association with Ludman. Eventually, with the realisation of the weight of the penalty for those involved in or associated with arson, they did talk. It emerged that Ludman had spent the evening drinking with others at the George and Dragon which he left at 11.20 p.m. before proceeding through the School grounds from the Howard's entrance in St Peter's Street. Ludman then, having separated from his associates, went his own way. One of them said he heard the sound of breaking glass behind him as he himself went on to the St Andrew's Road flat. Some time later – it would have been close on midnight – Ludman entered the flat and invited them to come out on to the fire-escape to see the sight. They saw the early flames leaping out of the School roof. Later they went down on the field and watched from close quarters.

After this there was no doubt in Superintendent Rogers's mind. The thing was to get hold of Ludman quickly. He knew his haunts. At 11.20 a.m. Ludman was found in the King's Arms and arrested. At 11.55 he was taken by the Superintendent to a house in Goldington where a number of items of stolen property and some cannabis were found. And after this it was Greyfriars Police Station and fuller questioning.

Superintendent Rogers took Ludman to Bedford School the next day, 23 March, to confront him with what he had done. He showed no remorse. He said, 'It's something to look at.'

On Monday, 23 July 1979, Ian Ludman was brought before His Honour Judge Robert Lymbury QC, accused on eleven counts of arson, burglary, obtaining by deception, and possession of cannabis. What follows is a summary of the proceedings at the Crown Court, Bedford.

The Prosecution stated that Ludman was interviewed twice. At the first interview he admitted the burglary and other counts since the evidence was undeniable; but the fire charge (count 8) he denied, saying, 'There is a little bit of history there, I would not like to see that go up.' When he was interviewed the second time he admitted responsibility for the fire, saying that he did it to destroy his fingerprints.

School life continues. *(Bedford County Press)*

Giving evidence on Ludman's background, Superintendent Rogers showed that he had never been able to hold down a steady job. Moreover, there was a record of ten previous convictions, one of them for the arson of an army jeep.

In his judgment Judge Lymbury said:

You pleaded guilty to count 8 on 3rd March, the fire at the Great Hall of Bedford School, which destroyed a building of architectural interest and the contents, portraits, books, records, 780 chairs, many of these donated by Old Boys and irreplaceable. This struck at the very heart of that school, but no doubt it will survive because it is made of sterner stuff than you can appreciate. The damage caused was approximately £2.5m and you watched the burning with glee.

You are void of social conscience and feelings for other people. The charge in count 2, the arson of 3rd February at 122 Howbury Street, causing £7,600 damage, was small by comparison, but more serious because you destroyed this woman's home. The charge in count 10 is arson of the Chicken Bar-B-Q in Bedford where the damage was £1,800. Each of the three offences of arson were committed in buildings in the town, where the fires might have spread and caused further damage and injured persons. But for the timely intervention of the Fire Brigade, people might have been injured and untold grief caused.

These are crimes of the utmost gravity and you have a previous conviction for arson. You are a social menace and there is a strong probability that you will remain a serious risk for many years. I am clear as to where my duty lies.

On counts 2, 8 and 10 (the arson charges) – imprisonment for life;
The burglary charges (counts 1, 3, 4, 5, 7 and 9) – 30 months' imprisonment;
Obtaining by deception (count 6) – 12 months' imprisonment;
Possession of cannabis (count 11) – 6 months' imprisonment.

All concurrent.

The work of the police, Superintendent Rogers and his team, in tracking Ludman and bringing him to face the consequences of his crimes was commended in court, and the then Chief Constable of Bedfordshire, Mr Anthony Armstrong QPM, paid his own tribute to the tenacity and professional ability, including skilful and determined interrogation, which had led to Ludman's conviction.

As a postscript to this side of the story it is interesting to note that from the time of Ludman's arrest there was a significant reduction in the number of fires in Bedford.

'As I was getting undressed I heard the sound of breaking glass coming from the direction of the school. Not one smash but rather a crackling and tinkling sound at irregular intervals. I pulled back the curtain and was confronted with a sight which, of the many from that evening, was to leave me with the most lasting impression. From the window I could not see the school building directly but, even at that early stage, the flames lit up the school fields so that I could see across to the hedge leading on to Park Avenue. I just stared out for a couple of seconds until it finally sunk in that the main building must be on fire.

There were as many people milling around that night as there was speculation as to how the fire had started. I remember Burnaby Road covered in heavy black hoses as the Fire Brigade pumped the drains for every drop of water they could obtain to spray onto the fire. The scene on the school field was rather like the one from "Close Encounters". People were standing around staring at the fire, their faces lit up and their mouths open. The heat from the fire was incredible. It was hot enough to warm our faces as we stood on the flat roof at Burnaby.

The next morning was an "exeat" Sunday and I still didn't believe it myself when I told my parents about the fire when they picked me up the same day. I suppose that is what shock is all about . . . your mind refusing to accept what your body has seen and experienced. The typical comment whenever I have had occasion to relate this story is one of every schoolboy's dream to see his school burnt down. But there really was no feeling of elation at the sight or consequences. It was a sensation of unreality which was to last for many months after the fire until one became accustomed to the new status quo . . . the pre-fabs and English in the pavilion.'

Hamish Ferguson

*   *   *

'When I arrived in the centre of Bedford the sky was illuminated by a bright yellow glow. When I reached the end of Burnaby Road I could see that the whole building was ablaze. It was the most fearful thing I have ever seen. There were over one hundred spectators, many of whom had no connection with the school. Several fire engines were there but had no control of the fire. After about half an hour I went home but was able to see the conflagration quite clearly from my house on Manton Heights. The fire was so powerful that I was surprised that the shell of the building was still standing. I was even more surprised when I discovered that my classroom (C13) was untouched as were the books of all the boys in my form. My two Russian dictionaries are next to me as I write this, neither the worse for the experience.'

Mr John Pidoux
Modern Languages Master

From *The Ousel*, Summer term 1979

Comments made, shortly after the fire, by eye-witnesses

. . . we rushed out and ran towards the school field. As we passed the Chapel there was an orange glow through the windows and this suggested to me that the Science Block was ablaze. We rounded the corner of the Chapel path and stopped horrified and trembling as we saw a pinnacle of flame about sixty feet high rise from the spire . . .

. . . there were five of us in the dorm' and we got up and went to the window and there was the most terrible sight I have ever seen in my life: the whole School from end to end was enveloped in one huge blaze. At first none of us could believe it . . .

. . . rows and rows of spectators were lit up on the rugger pitches by the roaring flames. Tiles began to fall off the roof, then suddenly the middle of the roof gave way and disappeared like water down a plug-hole . . .

. . . the sky was completely lit up, and enormous burning beams could be seen collapsing through the orange windows. The fire seemed so wild and monstrous . . .

. . . the collapse of the roof had moved the central spire to an angle of about 20 degrees towards the Dining Halls and about ten minutes later it too gave up its fight for survival and crashed to the floor of the inferno . . .

. . . the aim of the firemen was to save some of the classrooms in the East wing and to save the records and filing cabinets in the Head Master's study. Two firemen climbed into the study and at that moment the leaded windows of the Great Hall collapsed and shattered. One fireman had tried to control the blaze from above on an hydraulic ladder, but the heat was too great . . .

. . . from there I walked round to the south side just in time to see the language laboratory floor give way, and the momentum generated took all three floors into Mr Braggins's room . . .

. . . the biggest shock came the next morning. I was half expecting to wake up, look out of my window (in Kimbolton Road) and find that the building would be exactly as it always had been. But as I looked out I still saw the red walls, although large areas were blackened, and now the roof was non-existent, the chimney stuck comically out of the wreckage . . .

. . . the upper floor was a shambles with windows broken and roof beams swaying in the wind, small puffs of smoke curled upwards where the fire wasn't quite out, in one of the rooms curtains still hung from the windows . . .

. . . a smouldering mass of rubble surrounded by four walls. It was difficult to associate these walls with the banisters sticking awkwardly out of them with the beautifully wooden-panelled pride of the School . . .

. . . my first clear view was from Park Avenue . . . the roof and tower were gone and in their place was nothing. The School now had a completely different outline. The line of the roof had been replaced by the irregularity of the walls, with the chimney stacks adding still further to the effect. Light was shining through the north windows . . .

. . . the fire doors had done a very good job in containing the fire. On one side of these doors there was nothing but ashes. On the other the panelling is brightly polished and the notices on the noticeboard are untouched . . .

. . . I arrived back at my boarding house just after ten o'clock on Sunday evening. I walked into our television room to catch about twenty seconds of a filmstrip on television showing the burning inferno that was our School . . .

. . . the Old Bedfordians must have felt much worse emotionally to think that their old School did not exist any more. They do not know many (present) boys, so they are really only attached to the grounds and the building . . . all their proud moments in the Great Hall have disappeared . . .

Chapter III

Living in the wilderness

This may seem a journalistic caption to choose as a chapter heading. In fact it encapsulates the situation in which everyone found himself. The Head Master established a temporary headquarters in 6 Burnaby Road which on the first two floors incorporated the Bell Room and stationery store; an administrative centre with Head Master's study, and with the school secretaries; and an interview room. Perched up in the attic were Murray Fletcher, the Registrar, with the School Archivist, his wife, Pauline. The Staff Common Room was at the opposite end of the campus in what used to be the Music School Library. Assemblies were held in Chapel, and the new classroom block was a village of twenty-two temporary huts which sprang up like mushrooms just north of the new Biology wing, where the old 1st XI nets used to be.

Austin Hall portable buildings arrive. *(Bedford County Press)*

The Lower School was centred on another two temporary huts dumped outside Howard's which, together with the Art rooms and the Craig Building, provided just about enough accommodation. But the Lower School had a rough time of it. It must be remembered that during the whole of this period of twenty-nine months (seven terms) the School Estate was occupied – no other term is adequate – by several armies of builders. A number of firms combined in the restoration of the Main Building. They not only needed wooden huts as bases on the site, they also needed access: and the various vehicles large and small deposited along their routes vast quantities of mud. Add to this similar activity round the Recreation Centre which was being constructed at precisely the same time, and one can begin to imagine the unpleasantness underfoot for all who had to proceed to the extremities of the Estate. Those who suffered most were the masters and boys of the Lower School.

To put it mildly, the lines of communication were stretched. That in itself created problems. But the greatest problem of all, for everyone, during this period was the monstrous presence of the fire-scarred Main Building. Wherever one went that gaunt outline commanded attention. It loomed not only by its physical presence but by the power it inevitably exercised over one's thinking.

However, if in this sense the influence of the blackened building was oppressive, in another sense it provided an ever-present challenge. For some inexplicable reason the human spirit seems to take wing when things are desperate. Among Staff and boys throughout this long period of waiting, morale was extremely high. Everyone worked together in the rescue operation and sustained their efforts at a level well above normal. Moreover, in the early days of the Summer term the leaves of the vine around the building emerged from their winter hibernation and clothed the blackened walls with vivid green.

To describe the temporary classrooms as a village is reasonably accurate. First of all there was much debate as to the most suitable site. Out of six possible situations, an area on the field immediately west of the Chapel was selected. Rigid timetabling was essential if the village was to be in position and ready by the beginning of the Summer term. Everything that was needed had to be thought out in specific detail well before the end of the Easter term little more than a week away. The Bursar planned his campaign in meticulous detail. There was an appeal for volunteers, and squads of boys and masters, who were to provide the labour force were duly briefed. Financial transactions with Austin Hall were completed and dates fixed for the arrival of the prefabricated units. At the same time the Bursar completed arrangements with the LEA whose County Supplies Officer most generously made available to the school, from various depots around the county, literally all the furniture needed for seventeen of the classrooms. Furniture for the remaining five rooms was ordered from PEL Ltd. Arrangements for collection and distribution were, necessarily, complex.

This photograph emphasises the degree to which the unsupported brick gable ends are exposed. Their resistance to wind and weather is evidence of the craftmanship of the Victorian builders and the quality of the materials used. *(Arup Associates)*

Meanwhile, before the prefabs arrived, before the furniture was collected, the site had to be prepared. This meant ensuring that there was an adequate supply of electricity, and so a sub-station was located between the Chapel and the Music School, and trenches were dug for the electric cables. Then there was drainage to consider since the amount of rain water collecting on twenty-two flat roofs would cause extensive flooding if it could not get away. Connection was made with the rainwater drains by the Biology Block. Finally the village must be provided with streets for access to classrooms, and so a minor roadmaking operation was undertaken using paving slabs raised slightly above the level of the surrounding grass. These were so expertly laid that they remained firm and level throughout the period despite the weight, volume, and exuberance of young feet.

The interior of the Great Hall after clearing away the debris. *(John Laing plc)*

If 'village' is a fair description of the temporary classroom complex, 'mushroom-like' is an equally fair way of describing its sudden appearance on the field. The Easter term ended on Tuesday, 20 March: the first Austin Hall classroom appeared on 22 March. All the services indicated above – preparation of the site; drainage; electricity supply; roadmaking; collection, transport, and allocation of furniture into rooms; even the preparation and grass-seeding of certain areas – all were completed by the beginning of term on Wednesday, 19 April. Mushroom-like indeed: except that, perhaps, the word connotes appearance without effort.

The word 'effort' brings to mind another labour of this time. Staff and boys had, inevitably, lost books and other personal possessions in the blaze. To collect details, estimate their value and complete all the necessary paper work was a mammoth task, devoid of stimulation. In charge of it was Mrs M. Morgan, seconded from Messrs Thornton Baker, who coped nobly with these problems as well as with many other aspects of the insurance area.

This chapter makes no attempt to chronicle in detail the whole period during which the Main Building was restored. Rather it will attempt to give a selective impression of what it was like to be at the School as master or boy during this time.

First of all let it be clearly understood that this was no period of mere survival: of hanging on somehow and making do with half-life until full life could be resumed. Half measures were never considered by man or boy. All rose to the occasion and produced the best of which they were capable, surpassed themselves even.

One indication of this is the academic achievement of those whose chances might seem to have been imperilled by loss of revision notes and improvised teaching sites. Special circumstances forms were prepared for every boy taking GCE 'A' and 'O' Level exams. This meant making a careful assessment in each case of likely performance: no small task. The Oxford and Cambridge Board were most co-operative. In the event, however, these forms proved merely to be a precaution in case of disaster: the boys excelled themselves. Reviewing the examination results at Prize Giving later in the year, the Head Master said that, while he was conscious of the danger of using statistics, it was impossible to deny the achievement of an 86 per cent pass rate at 'A' Level with 41 per cent of the passes in grades A and B: and at 'O' Level, the achievement of an average pass rate of 8.6 passes at grades A, B, or C for each of the 160 candidates.

Perhaps these excellent results were a product of the stimulus of truly adverse examination conditions. The examination room was a bleak concrete acre at the very top of the Harpur Centre. The floor was cement through which square pillars thrust at intervals to sustain the concrete roof. There was a remarkable echo which intensified the squeal of chair or desk on concrete. The atmosphere was claustrophobic with the windows shut, but with them open the noises of drilling and banging from a site being excavated nearby were more than distracting. To this arena the victims made their way daily, walking from School, or from home, or from their boarding houses. Spirits throughout were remarkably high. There was a tremendous sense of adventure.

Nor was the adventure, so far as the staff was concerned, without its lighter side. Stemming largely from his sense of fun and the lively sparkle he touches off in others, Mr Briggs began a file to contain the jottings made by masters during invigilation. As they perambulate, keeping one eye open for (unlikely) breaches of examination rules, masters, directing their thoughts towards some of the droller incidents and implications of school life, may, without any slackening of vigilance, make notes, jottings, or compose verses. Many of these ended up in Mr Briggs's collection. (*As shown on page 70*)

Examination results were not the only indication that school life was continuing in all its fullness. The cricket XI had its most successful season on record, while the oarsmen performed very creditably at every level and the young 1st VIII put up a tremendous showing at Henley. Sport of every kind flourished; Mr Barlow staged *Twelfth Night* in the garden of 39 De Parys Avenue; and the Festival of Music was held as usual but, by kind permission of the Headmistress, in the Dame Alice School hall.

Moreover, as a further proof of continued development, the Christmas term 1979 saw the inauguration of a new venture, the opening of the Permaprene all-weather pitch on the Gordon field. With hindsight it can be said that this pitch has made an immense contribution to the standard of hockey throughout the School. As a pitch it is fast and true, it drains instantly, and only frost can put it out of action. In the summer it translates into twelve excellent tennis courts. At the time of writing it is true to say that no School fixtures in tennis or hockey have had to be cancelled.

Also in 1979 Bedford High School Hall was generously lent by its Headmistress to house both the School Prize Giving and the School Play, the latter a joint High School/Bedford School presentation of *The Crucible* directed by Mrs Goddard. The standard of acting was high and the performance as a whole genuinely moving.

At Prize Giving the prizes were presented by Sir Peter Parker, then Chairman of British Rail and himself an Old Bedfordian. Sir Peter spoke with passion on the need for imaginative management in industry. He concluded, referring to Bedford School, 'That fire, for all its devastation, lit up again, even in its tormenting blaze, the true sense of community and service that Bedford School means, and has meant to so many of us lucky enough to experience it. The Bedford Eagle rises, like a Phoenix, from the ashes. This management of change will be a glowing inspiration on the School record: it was something of a miracle.' Never was truer word spoken.

The end of the Christmas term signalled what was supposed to be the beginning of the actual work of restoration. But, to the casual observer, there was not much to see. At the end of March 1979 the Main Building had

Invigilation: Harpur Centre

Poems by members of staff written during G.C.E. examinations.

As white shirts crouch in a narrow sandwich
Spread two hundred feet above the town of Bedford,
While immediately beneath, a layer of Harpur Charity
Filters the intake of ground floor business
Into the minds of those who now above them,
Supported by the economic know-how
Set down on white examination paper
The iceberg tip of Godliness and good learning,
Would not Sir William's spirit rest in peace
To see this late twentieth century mini-Holborn,
This modern distillation of commerce and education
Finally united in his own Memorial Centre?   D.H.B.

There's a hubbub round the door
Of Harpur Centre's second floor
   When the 'A' Level's shortly to begin
Screaming 'Antony and Cleo,
Macduff, Donalbain and Fle . . .' – Oh!
   I simply cannot think with all this din.   A.E.B.

Rigor mortis set in at half past three.
Invigilator pole-axed: R.I.P.   P.D.B.

Less smell of feet this year!
I wonder why?
Smells don't escape from noise!
Perhaps they do?

This chance discovery
Could revolutionise
Laundry.   A.D.G.

   Where have all the flowers gone . . .?

A thousand young biologists, labouring through the hours,
Artfully wielding razor-blades to desecrate wild flowers.
A once-resplendent buttercup falls sectioned to the floor,
Its petals scattered, 'bias' oozing out through every pore.
Is this what 'logos' had in mind for 'Homo sapiens',
A dreamer's eye, inadequate, bunged up with plastic lens?
Has Eden's gardener then forgot his sovereign lordly role?
The guardian of creation; can he thus have sold his soul?
Ay, hard-hearted, disdainful does he Eden's portals slam;
For England's youth is writing a Biology exam.   T.A.R.

been fenced off and during the summer there had been the excitement of seeing cranes place Portakabins one on top of the other within the perimeter. There was a perpetual coming and going of white, blue and yellow helmeted figures. Vehicles of various sorts also came and went. But it was not until September that the Building was cleared of burnt fragments, charred timbers, metal strips, fallen ceilings, general mess and rubble. And if six months seems an excessive time for clearing up, one has only to look at the pictures to realise the size of the task. It would have been accomplished more quickly if the policy had been to knock the School down altogether. Then bulldozers and huge lorries would have been used. As it was, the whole operation had to be done by hand.

Even after the Building had been cleared, nothing much seemed to actually happen. The gaunt wreck just sat there magnetising the eyes of all who passed near. In fact, during all this time innumerable activities were taking place. The Staff Planning Committee was composing a detailed brief of what the new School should contain. The Architects were looking at the problems from every angle in an attempt to achieve, within the limitations of the standing walls, the extra space demanded by that brief. They came up with two proposals: raising the Great Hall and utilising roof space. Innumerable plans and drawings were made. However, to the outsider none of this could be known.

What was known, and, by clang and clatter and mud made its presence felt, was the development of the old gymnasium area into a Recreation

The skeleton of the new Sports Hall. *(M. J. Norris)*

71

Centre designed by Mr R. S. Hollins of Framlingham to house an indoor swimming pool, four squash courts, and an enormous sports hall. The gym itself was to be converted into a theatre. At the time of the fire it was a brave decision by the Governors to continue with this project. In retrospect it was the right decision keenly supported by Old Boys led by Sir John Howard, President of the Old Bedfordians' Club. Bedford School today would be the poorer without it.

Through the summer and winter of 1979 and the spring of 1980 the Recreation Centre gradually took shape. First the old Swimming Pool was filled in with rubble; then concrete verticals reached skyward. Mr Hollins's technique was to get the roof on first and fill in afterwards. It was fascinating to watch.

A poem written at the time brings back memories of what it was like to be teaching in a temporary classroom looking out over the school field towards all this activity.

The catwalk provides access to the nuts and bolts which secure the skeleton of the roof. The vertical lines at the top of the picture connect with the crane which has just lowered one rib into position. A later picture
(John Laing plc)

Here on blank paper record the scene.

School field with football posts, bare trees and the gardens
of houses in a residential road.

Over the grass a tractor, all wheels and colour, plays
truant, while the serious groundsmen bend.

Nearer, in this room,
boys are behaving like boys, most of them working,
losing or finding themselves in a world of words.

South, beyond the Memorial Library
where builders wade in mud, there pokes above
the roof line a small, yellow, blind
head, to an accompaniment of
prehistoric growling; then,
with a roar of sheer power
an awkward monster rears above the buildings
extending unimaginable lengths
of tubular steel neck,
gigantic, higher and higher,
throatily asserting.
A yellow mouth dribbles metal strings
which suddenly quiver, clamp tight
and hum the weight of the great concrete span
which, beneath, rises; barely swings.

Around,
a few bored men speculate
as, with immaculate skill,
with eye and judging hand,
a cased mahout
lighting a cigarette – no test –
guides machine and concrete tons until ends meet
and slack wires rest.

A boy raises his hand.
'Please sir, does it matter if it rhymes?'

To continue with a brief account of the completing of the Recreation
Centre is to leap ahead in time from the early months of 1980 to the summer,
when for the first time the 'A' and 'O' Level exams were held in the Sports
Hall: and then again on to 19 September when, complete except for
finishing touches to the theatre, the complex was officially opened by Mr
R. E. G. Jeeps, CBE, OBM, Chairman of the Sports Council. Thus was concluded a project which had been explored for at least five years before the
laying of the foundation stone in July 1979 by Mr D. P. Rogers, OBE (OB),

Sir John unveils the stone. From left to right: Mr F. Fleming, Managing Director of John Laing; A. C. W. Abrahams, Chairman of the Harpur Trust; Sir Philip Dowson, Senior Partner of Arup Associates; Mr C. I. M. Jones, Head Master; Mr H. P. Shallard, Chairman of the Bedford School Committee; Sir John Howard, President of the Old Bedfordians Club. *(Arup Associates)*

then Chairman of the England Rugby Selection Committee. It was made possible by generous subscribers who raised more than half a million pounds.

To return now from the completed Recreation Centre to the Main School Building, it has been said that despite the presence of building huts and a helmeted work force, nothing much could be seen happening. However, a Development Bulletin issued with the March *Ousel* published the finished Arup plans for the whole building. From then on things did begin to happen and progress could be measured by eye. The exterior walls were racked with scaffolding and assorted ladders. From breakfast to tea time, Saturdays included, there was a continuous banging coupled with the not-always-polite cries of workmen. The familiar scarred and charred outline was still a reminder of the disaster but the process of renewal was evident and gave a lift to the spirit. This was one year on.

Most of the unspectacular work had now been done. The sub-floor had been strengthened, ducted to contain a labyrinth of pipes and tubes for fresh air, for heating, and for such modern amenities as TV terminals, telephone cables and computer links to enable each department in the school to keep in touch with each other and with the modern world. By the end of the summer the whole of the ground floor was roofed over at first-floor level and the rooms below were taking shape. The new galleries on the first and second floors were in position and it was possible once more to walk round the Great Hall. Work was proceeding, too, on the rebuilding of the tracery windows.

It was at this time, summer 1980, that the Commemorative Stone pictured opposite was unveiled. The following account is taken from the *Ousel* Editorial.

On the final Saturday of last Term a simple ceremony saw the unveiling by Sir John Howard of a stone tablet which commemorates the restoration of the Main Building. Held on the floor of the Great Hall the ceremony had, as far as the setting would allow, an intimate and informal atmosphere. There were no rows of seats, no procession of dignitaries; a sequence of simple speeches given before an audience that stood informally on the unadorned concrete floor of the new Great Hall, unglazed windows behind them, the open sky above.

Listening to the speeches and watching the unveiling one was aware of many things: of the craftsman's interest in the enormous task of rebuilding – Mr Fleming, representing Laing Management Contracting Ltd, spoke of the task as one of the most rewarding he had personally come across; of the artistic and professional integrity of the architects, represented by Sir Philip Dowson, who had shown uncanny skill in interpreting the School's material and spiritual needs; of the immense amount of care and effort so freely given to the School by the School Committee, and in particular by Mr Abrahams and

The interior of the hall during the period of reconstruction. The scaffolding supports the many catwalks needed to secure the roof. *(John Laing plc)*

Mr Shallard; but above all of the overwhelming sense of relief and rejoicing that real progress towards the final re-occupation of the Main Building was being made and that it had been recorded by the unveiling of this dignified stone.

Inconvenience is a poor word to describe what the School is suffering at the moment. Building machines, ranging from vast crane lorries to diesel dumper trucks, manoeuvre their way round the School estate between the Wells Building and the Phillpotts Gates; the asphalt road-way gapes muddy craters, trenches lie open with half laid gas-pipes, wire netting surrounds the building compound, these are the order of the day. All the product of one chance lunatic action.

The *Ousel* Editor chronicles the inconvenience of living and schooling with all this work in progress. The unveiling of the stone gave morale a much-needed boost, indicating that the turmoil would one day end. The fire had dealt its blow but the School had kept going and completed the lap of one academic year with honour. That was 1979. Now in 1980 the half-way mark had been passed. True, another long year of endurance lay ahead, but the end could be seen.

The Grand Summer Ball, held on the night preceding the above ceremony, gave its own fillip to morale by providing a celebration of a different sort. It proved an attraction not only to those who attended but to everyone on the School Estate, for on the field immediately north of the Main Building was erected an astonishingly large marquee where 1,200 people set care aside for half the night. Inside it was like fairyland. Music was provided by Joe Loss himself, making a number of appearances during the evening, each time in a different, splendid costume; a charismatic figure indeed. The School dance band which opened the proceedings was at first taken to be the Joe Loss orchestra, so lively were the rhythms, until, on closer inspection, some well-known School personalities were seen behind the instruments. At midnight the Head Master was presented with a monster cheque by the Friends of Bedford School who had organised the Ball to raise funds. The total amount contributed was £3,500, which was used to help pay for the organ in the Great Hall.

In its final year the restoration seemed to make haste slowly. Slowly because the growth was on such a broad front. The Main Building was, of course, the focus of all eyes; and of all ears too. The sound of work in progress continued literally all day and all night. At night the noise was that of a machine which draws air through a plant which, by extracting moisture, decreases the time needed for building materials to dry out, and therefore speeds up operations. By day the noise was that of men at work. There were men everywhere: men moving to construct the roof; men on the north tower, continually, it seemed, hoisting aloft on a toy crane bricks and barrows of slopping cement; men inside the building, hatted against fall-out, with papers in their hands, pointing; men drilling holes; men fixing pipes and

The spire about to be lifted into position *(M. J. Norris)*

wires; men hammering, men shouting, men singing. Through the windows of the Arup and Laing huts men seemed to be holding perpetual conference. Things were really moving.

One major excitement was the arrival of the first lorry carrying the wooden skeleton of the roof. Then came a giant crane which bedded down on the tarmac and filled the air with diesel decibels as it hoisted each section aloft. Up on the roofline figures pointed and signalled. When the correct positioning was achieved they could be seen making final adjustments before bolting the sections together. The process was fascinating but it went on for a very long time. It is, after all, a very big roof. After that, possibly even more fascinating, was the arrival of the shaped cedar panels which, as they were lowered into position, exactly matched the configurations of the skeleton roof below them. Of course this was how it should be, but to any DIY enthusiasts watching (whose amateur cut-outs do not always match perfectly) it was a marvel to see precision on this scale. This operation, too, was inevitably a long one, but eventually it ended and the School, at last, had a roof again. Of a sort; for this was only the inner roof. Yet to come were the insulation and the felt, the battens and the tiles.

At Prize Giving in the Christmas term 1980 (once again held in the High School by the generosity of the Headmistress) Sir Philip Dowson, Chief Architect of Arup Associates, was the Guest of Honour. After giving away the prizes he made a sensitive and thought-provoking speech reflecting his vision of an architecture at the service of humanity. 'We must recover a comprehensive vision of the wholeness of the environment in which we live. In our mechanised society we should passionately emphasise we are still a world of men, and that man in his natural environment must still be the

A stonemason tidying up an almost completed window. Most of the tracery had to be renewed. Fortunately the original Robins designs were available from the archives of the Harpur Trust. All the stone used in the reconstruction came from the quarry which supplied the builders in 1891. *(John Laing plc)*

The School racked with scaffolding *(Alistair Muir)*

focus of all planning and design.' This was illuminating and struck home. How often in every town and every city one looks around and sees buildings which, though they may be brilliant, are essentially brutal because the human dimension is missing. Particularly good was it that Sir Philip should make this statement in a school, in the context of education; and make explicit the humanity of his approach to an audience concerned with the next generation, before representatives of that generation. His concluding remarks pointed the nature of the challenge that the technological society poses.

If we are to witness a search for a more human architecture, and one that seeks to develop methods that can compete with the pressures of scale and technology and to *revalue* them in human terms, then those of you who will become involved in this field have a massive task ahead – which must mean a closer understanding and integration between *all* the design professions to meet this challenge. That is, if we *are* to secure the ascendance of humane ideas, which, while of course embodying the technical, will not continue to do so at such great social expense.

The great wheel of time and of fortune turned, ushering in the year 1981 and with it fresh hope and the prospect of the restoration being completed on time. In the early weeks of the year very many things were happening. Because of the size of the building it was possible, by skilful organisation, to employ at the same time plasterers and decorators, electricians and heating engineers, stone masons, carpenters, and men actually putting down finished floorings. Some windows were being glazed while others were still being constructed. Also the outside of the building was being cleaned. Thought was now given to the positioning of car parks, traffic circulation and the ultimate landscaping of the Estate immediately surrounding the Main Building. The most dramatic development was the arrival of the new spire, made all in one piece, standing forty feet high, and hoisted by crane into position where it awaited only the finial to be complete.

'Topping Out' was scheduled for 29 April. 'Topping Out' ceremonies, which are observed all over the world, indicate that a building is fully roofed and therefore watertight; a significant achievement celebrated usually by a party given for the builders by the grateful client. The highlights of this particular 'Topping Out' were to be the placing by the Head Master of the finial – in this case a crown of flame wrought in copper – on the pinnacle of the spire; and the unfurling of the School flag by Mr A. C. W. Abrahams.

There was considerable press and TV coverage, in part because of the widespread interest in the fire itself and in the subsequent restoration process; in part because the School had prepared a 'time-capsule' to be placed in the spire awaiting the year AD 2552 when Bedford School will be a thousand years old. The contents, which were intended to give some idea of school life in 1981, included a prize-winning essay by a fourteen-year-old boy, Nicholas Tinworth, on what he imagined life would be like at Bedford School when the capsule was opened. The capsule, of black fibreglass and made in the School Workshops, had been immersed by Texas Instruments in pure dry hydrogen and then hermetically sealed.

By three o'clock a large crowd on the field watched as the Head Master, Sir Maurice Laing, Mr H. P. Shallard and Nicholas Tinworth were hoisted to the heights of the spire by a bright red crane known as a 'moon shot'. On terra firma again with the spire duly crowned and the 'time capsule' lodged, Sir Maurice Laing, the Head Master and Mr Shallard spoke briefly on the significance of the event.

They were followed on the crane platform by Mr T. W. Fleming, Managing Director of Laing Management Contracting Ltd; Mr A. C. W. Abrahams; Mr M. E. Barlen; and Glenn Milne, Head of School. Before take-off Mr Abrahams publicly invited Glenn Milne to share with him the privilege of unfurling the School flag. This, at tower level, they jointly did; and the crowd of boys, parents, staff, old boys and well-wishers were evidently stirred as the flag was unfurled and the eagle spread its wings. At ground level again Mr Fleming and Mr Abrahams added their tribute of a few words before the crowd dispersed.

After this, guests were invited to the Great Hall by Laings, hosts on this occasion, to meet the men who were doing the actual building. It strongly emerged that all took a very considerable pride in the quality of their work.

The Time Capsule

The Capsule, which was made of fibre-glass by Mr R. F. Goodacre in the School Workshops, contained the following articles:

Prize winning essay on Bedford School 2552.

Bedford School Prospectus containing lists of Governors and Staff as at Easter 1981, and details of fees, etc.

Development Bulletin No. 3.

The 1980/81 School list and Upper School timetable.

Ousels of June 1979 (the Fire) and March 1981.

Ousel Calendar Summer 1981.

Detention form.

School Rules.

Bedford School Chapel Service Book.

Bedford School Restoration Brochure.

Aerial photograph of the shell of the Main Building.

Facsimile of John Tavener's composition *Risen* commissioned for Bedford School Restoration Year Concert.

A selection of Oxford and Cambridge Examinations Board papers: including 'O' and 'A' level papers in English, French, History, Mathematics and Physics (Nuffield), Chemistry (Nuffield), Biology 'A',

Wrappers of current Tobler-Suchard products.

A video tape from Unilever Ltd indicating the nature and range of their scientific research and its value in social terms.

Two of the latest Texas Instruments micro-processors.

A School video tape of Bedford, viewing: the Town Bridge and Suspension Bridge; St Paul's Church; the Blore façade and the Harpur Centre; County Hall; the Railway Station; traffic from buses to bicycles; Bedford School and the Chapel.

Before their ascent to the spire the Head Master and Nicholas Tinworth pose for Philip Tam, a boy at the school. The Head Master is holding the finial, a crown of flame wrought in copper, destined for the pinnacle of the spire. Nicholas Tinworth holds the Time Capsule which, among other things, contains his prize-winning vision of Bedford School in 2552 A.D., a thousand years after its foundation.

On the way up. From left to right: Mr H. P. Shallard; the Head Master; Sir Maurice Laing; Nicholas Tinworth. *(John Laing plc)*

A street plan of Bedford.
A boy's tie (Ashburnham) donated by Beagleys.
A box of chalk.
Copies of the *Bedfordshire Times* and *The Times*.
Miniaturised copies of the *Guardian* and the *Daily Telegraph*.
Pages from *The Times* and the *Telegraph* indicating the beginning of Summer term 1981.
7 slides (photographs) from Arups.
1 miniature yellow helmet (pencil sharpener) from Laings.
Minutes of the 50th Staff Planning Committee Meeting – with Arups.

PRIZE-WINNING ESSAY

Bedford School AD 2552

The aged building sleeps in silence. Slowly the warm glow of sunrise creeps across the horizon. Suddenly and silently banks of tiles on the roof of Bedford School slide back and large solar panels are revealed. They begin tracking the sun on its celestial path across the sky. A faint mist appears to cover the building; in reality an air-tight skin of 'atmo-plastic'. Three hundred years before the present time, man finally destroyed the earth's ozone layer and now harmful ultra-violet rays scorch the earth. A century later, another man-induced disaster – the release of poisonous gases into the atmosphere – led to the destruction of nearly all plants and almost destroyed the world's economy. The 'grass' which surrounds the old building is manufactured from chemicals. The real trees that survived are encased in ballons of 'atmo-plastic' to protect them from the deadly rays and gases, as the School itself is protected and the people within.

Soft humming sounds fill the air. Small bubble-shaped hovercraft appear, each containing one person. A computer in each communal home is programmed to direct each vehicle where the occupant (or his or her parents) want it to go. The City's main computer controls the magnetic roadlinks along which the vehicles travel. There are no traffic-jams or accidents. Solar energy is used.

The 'bubblecrafts' pass through air-locks doors to the decontamination area where the occupants descend to be sprayed with neutralising chemicals and then move through into the School, while their vehicles are automatically parked high in the air outside, under magnetic shields. Now the children are within the world's oldest and best preserved School still in existence.

They look very similar to children in photographs saved from about the year 2000 AD. They wear a dark blue uniform, simple jacket and trousers, but in disposable stretch material. Boys and girls wear the same uniform. The masks they wear constantly elsewhere are not needed here as air is sucked through purifying and oxygenating devices and circulated throughout the building.

Only children with a very high science potential are accepted for this School and many leave to take charge of the construction of new space colonies. Those whose parents are already working in 'near-space', board with a community family in the City. They will only join their families when their education is complete.

At 9.00 a.m. flashing lights signal the start of classes – fifteen minute sessions on Maths, Chemistry, Physics, Biology, Astronomical Science, World-English, Space Dialects and Space Geography.

The learning process is simple. Earphones are plugged in as each child sits before a small screen. The lesson appears – a mixture of words, symbols and pictures, plus sound from the 'Telephotronic system'. A central computer, 'The Brain', controls all teaching programmes for all ages and subjects. Lessons are learned almost effortlessly by a combination of waves and sounds pulsating on the brain, which permanently fix the information there. Achievements are somewhat varied because the child must contribute 'concentration', and some achieve this better than others.

Teachers are still needed, for children continue to be as well or as badly behaved as they have always been. Their teachers enforce discipline and encourage original thought and discussion.

Optional subjects are ancient languages: Latin, Greek, French and German. Divinity may also be studied – mainly Christianity. The world's religions died out almost completely during the period 2300–2400, but in the past one hundred years Christianity in particular has been gaining ground again, possibly because mankind has not yet discovered how to prevent death. The School gathers together in the Great Hall each day at noon to offer prayers to the invisible God who rules the Universe.

A very popular extra subject is Art. Not the flat, unmoving pictures of long ago, but schemes and designs formed in the mind, which by concentration and the laws of physics are made visible in the air or against a suitable surface, where they appear to live and move.

The Ancient Music Club performs on antique instruments such as pianos, violins, electric guitars, synthesisers, etc., playing old classical, rock, pop, jazz and punk music. The Modern Music assembly is vast, using the latest sensorywave instruments for Blackhole Fluke music.

Following Assembly, there is a clattering of feet and a chattering of voices as everyone moves to the dining halls for the favourite relaxation – food. All sit at long tables, each place laid with a plate, knife, fork and spoon and alongside are two buttons. When the red button is pressed 'thought-rays' are activated. Then everyone concentrates hard, picturing in their minds the exact meal they would like to eat and it appears immediately on their plates. It has the appropriate flavour and tantalizing cooking smells, but this is only an illusion. In reality the meal is merely a combination of the vitamins, protein and carbohydrates need for healthy living. Many years ago it was found that people would not eat meals reduced to tablet form, although this saved vital energy and production resources. This modern method was devised which has proved most successful. The action of the green button is to destroy the disposable plates and cutlery.

Each afternoon everyone is free to follow their own favourite occupation. The ancient games of rugby and hockey still continue, but in different form from that played several centuries ago. Everyone now wears a protective suit and mask. The movements called 'tackling' and 'scrums' are forbidden, and hockey is played by moving the ball with hands and feet, not a thick stick as in olden times. The size of pitches has had to be increased, for modern children have an average height of 2.5–3 metres when fully grown.

Only senior children are allowed to use the Library, which has a collection of almost 500 books. As their use has been replaced by video-cassettes, very few books are now left in the world and therefore these are highly prized.

The end of another day. The solar panels have tracked the blistering sun on its 180° journey across the sky. Now twilight spreads its cooling fingers over the misty buildings and synthetic grass, covering them with a black veil. Children, teachers, bubblecars – all have gone. The only sound is a faint whirr as the solar panels sink back into the roof and the click of tiles sliding into place. The great building, crowned by its triumphant spire, sleeps calm and undisturbed – just as it has done for over five hundred years. Is this 2500, or 2000, or 3000 AD? Who can tell?

Nicholas E. Tinworth
April 1981

The long haul was nearly over. On the final Saturday of the Summer term visitors were allowed to explore the building and see for themselves what had been done. Nothing was actually completed at this point, the middle of July, and there was a litter of oddments and mess of one sort or another: wood shavings; loose wires; unattached radiators; pieces of shaped stone awaiting placement; boxes of quarry tiles; brushes and canisters of paint; workmen's tools; and much more besides. Visitors marvelled at what had been done but questioned whether the multitude of loose ends could possibly be tied up in the month or so remaining. Moreover the Copeman-Hart organ had to be 'voiced', and for that complete silence was required. It was conjectured that Mr Hart would have to 'voice' all night and sleep all day.

There is no need here to spin out the final details. Suffice it to say that the work was completed and the building handed over to the School in time to begin the new academic year. It was with a sense of wonder and rejoicing that boys and masters gathered in the new Great Hall for the first assembly of the Christmas term.

Chapter IV

The Replanning Process

The eruption of flame that scattered the School's heritage in clouds of spark and ash across the night sky was to draw into its incandescent vortex the ideas and aspirations of many generations. Plans formulated in embryo during the previous decade, studies projected but unrealised, hopes implicit but still undefined, all surfaced to demand immediate attention. The fire focused the mind upon instant needs and long-term requirements. The time factor pressed molten thought into practical and structured moulds. A surge of concentrated and creative thinking absorbed Governors, Staff, Parents, Old Boys and Friends of the School. This was to spill over on to committees, architects, officials, builders and many sectors of the School community. They were fortunate to find at hand the men and materials to realise so many of their hopes.

The Head Master was thinking ahead even at the height of the inferno. By 4.00 a.m. he had summoned an emergency committee to meet in his house at 10.00 a.m. the same day. For those who assembled on that fine and sunny Sunday morning it was difficult to grasp the implications of the previous night, and only the occasional glance back across the School playing fields at the still smouldering ruins brought home any realisation of the full extent of the disaster. Calmly and confidently the Head allotted to each his special role. He himself would be writing to parents and meeting the insurers with the Bursar and Trust Officers in a few hours' time; his Secretary would be moving what was left of the School Offices into 6 Burnaby Road, which had only recently been purchased for the School and was still empty; the Usher would look for alternative classroom space wherever it was to be found – in boarding houses, pavilions, laboratories and dining halls; staff were allotted to deal with burnt books, examination notes, records, reports, offers of help and letters of sympathy. The Staff Common Room could move into the Music School Library; the Chapel would be used for morning assemblies. Special arrangements were put in hand for 'O' and 'A' Level examinations. School was to function 'normally' until the end of term; not one teaching period was to be lost.

Crises create their own anomalies. At the meeting there was a man in a dark suit. He was assumed to have something to do with the fire; but when he left no one quite knew who he was or why he had come! The Chairman of the School Committee, Mr H.P. Shallard, arrived half-way through the morning in a hurriedly hired car; his own was undergoing repairs. Other Governors came in to view the scene and to offer their sympathy and support. The Head Master, Bursar and Trust Officers had their first meeting with the insurers later that morning: their agents sounded reassuring, anxious to mitigate the initial shock. The Chairman of Governors, Mr A.C.W. Abrahams, arrived on the Monday. Amid sorting and sifting, the start of a new School day in improvised surroundings, random thought and purposeful gesture, future guidelines began to emerge. By the time the Head Master addressed the Staff at the end of the day, focal points had crystallised. These were confirmed twenty-four hours later in an address from the Chairman; they were to point the general course of the re-planning and restoration.

The immediate objectives were clear enough. The continued functioning of the School had been achieved, but purpose-built temporary accommodation would have to be ordered, sited and erected before the beginning of the coming term. The burnt-out ruins had to be made safe from wandering interlopers and souvenir hunters. The extent of the losses to the School and to individuals had quickly to be quantified so that negotiations with the insurers could begin. As for the future, the Main School Building was to be restored or rebuilt by September 1981. Two and a half years was the most, the Head Master believed, that a school could survive in temporary buildings without suffering academic or social damage. But the two and a half year target was to impose some significant constraints. The building had only recently been listed as of Grade II architectural interest. Once the outer walls had been declared safe and capable of load bearing there would be pressure from conservationists and Old Boys to preserve the façade. Demolition would require consent and take up valuable time; planning a new building to suit might take even longer. Time would be pressing while inflation appeared to be rising. Thus restoration within the shell of the old building became the likely outcome.

It was on the Tuesday that the Chairman had been moved to declare that the educational brief for the restoration would be prepared by the Head Master and his Staff. He was indeed the Staff's nominated Governor to the Trust, but it was in any case his belief that they would best know what to include in an up-to-date educational brief. Pressure of time would, however, leave little room for much fundamental thinking. Perhaps it was as well not to be tempted towards too much originality. Existing ideas and blue-prints would have to be used. There were indeed an unexpected number available: plans and feasibility studies going back some ten years lay suspended in various files and cabinets. These, carefully scrutinised, sifted and tested, were to become the source of much of the future planning and debate.

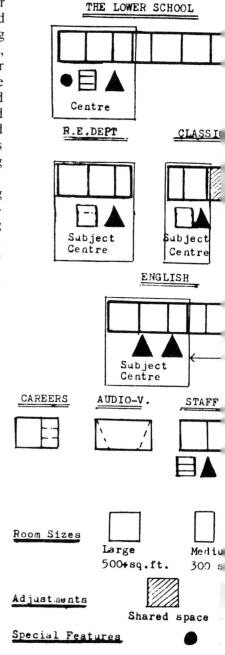

The ideal space distribution
(Staff Planning Committee Report)

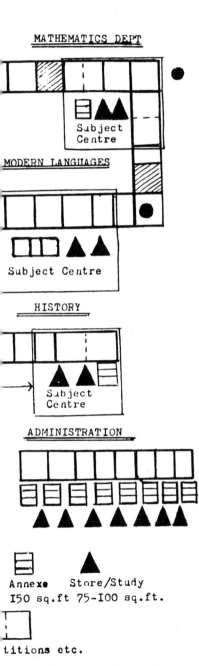

MATHEMATICS DEPT

Subject
Centre

MODERN LANGUAGES

Subject Centre

HISTORY

Subject
Centre

ADMINISTRATION

Annexe   Store/Study
150 sq.ft 75-100 sq.ft.

...titions etc.

...ut rs, Language Laboratory

Meanwhile, the Governors set in motion the machinery to ensure immediate objectives and point the way towards the fulfilment of long-term goals. Committees, so often the butt of criticism for their procedural delays, can in an emergency be inspired or reconstituted to promote rapid and effective action. Here was to be no exception. Before the week was out the Chairman had explained the position to those concerned and a new Restoration Committee of senior Governors was set up to cover the immediate crisis (see Appendix). Their roles covered security, insurance, temporary accommodation and the appointment of an architect to carry out the reconstruction. The Surveyor to the Trust worked day and night to ensure the safety and stability of the ruined structure. This meant, in effect, that while boys had to be able to get round the building they had to be prevented from getting into it! Scaffolding, fencing, covered passageways soon provided protection from loose tiles and listing gables. Insurance negotiations were to be more protracted and less exposed to public view. First impressions had been too optimistic. Negotiations would follow the usual course with a period of skilful but friendly bargaining leading to a reasonable and balanced compromise. While the Bursar dealt with the many claims from members of the School, particular Governors were delegated to handle the insurers and loss adjusters and to bring in additional advisers to strengthen their legal position.

A small sub-committee of Governors and Trust Officers was nominated to deal with the temporary classroom accommodation. This was needed for the start of the following term so, as the Clerk noted laconically, time was of the essence. Sir John Howard was drafted in to help guarantee success. Rapid visits by a combination of Head Master, Surveyor, Bursar and Usher to University College School, also recently hit by fire, and to view various prefabricated units led to an order for twenty-two Austin Hall classrooms with cloakroom and storage accommodation to be erected within four weeks on a site laid out between the Chapel and Biology Laboratories. The Contractors arrived on site on the last day of term. In spite of a wet spring, mud and the need for massive machinery, the sections were assembled and completed by 17 April. The area around the Music School garden, with the Chapel and Science Block cradling the temporary classrooms, was to become an educational oasis among the building sites that came to dominate the heart of the School Estate.

The appointment of an architect to deal with the evaluation and reconstruction of the area was clearly of the utmost urgency. The Restoration Committee had decided at their first meeting to make an initial approach to Dr Bernard Feilden, Old Bedfordian, Surveyor to York Minster, architect of St Paul's new school, who had already drafted two feasibility studies on the development of the School and its Estate during the previous eight years. Unfortunately his job as Director of the International Centre for Conservation in Rome did not allow him the necessary time, but during a lengthy telephone call with the Chairman on 12 March he did suggest the name of his friend Philip Dowson, Senior Partner of Arup Associates. A quick dip into *Who's Who* revealed that both Architect and Chairman shared the same Club! An evening at the Garrick and the first links were joined: Arup Associates would make a presentation to a full meeting of Governors on 17 March at the end of that week. A second presentation would be made by Frederick Gibberd and Partners who had recently completed work on the Harpur Centre, behind the Blore façade of the old Modern School, and had a record of distinguished achievement. Both teams viewed the ruins on the day before the presentations were to be made. Those who escorted Philip Dowson were quickly struck by his critical discernment and personal charm. His comments so often seemed to strike a responsive chord. At the presentation to the Governors this sense of affinity with the School and its problem appeared to be confirmed. After both presentations the Governing Body went into Committee and a first ballot showed a decisive majority in favour of the Arup team. Arup Associates were thereupon unanimously invited to undertake the job of restoring the School.

19 March, two weeks after the fire, brought term to a close. The buildings had been made secure, temporary accommodation had been ordered, negotiations with insurers had begun and architects had been appointed. Long-term planning had now to follow with the several parties and agents involved being effectively drawn together. The Governors, therefore, completed their preparatory work by drafting a plan of operation and redefining their committee structure. As immediate pressures eased so the need for an emergency Restoration Committee appeared to diminish. After three momentous meetings its functions and the future surveillance of the planning and reconstruction process were to revert to the School Committee – those Governors immediately responsible for the running of the School – whose Chairman, Mr H. P. Shallard, was to play a leading role during the next two years.

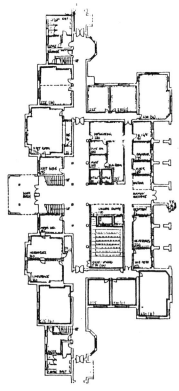

Ground floor first draft (Plan A)

The Architects were now to liaise with a Staff Planning Committee which had been appointed by the Head Master and a new Project Sub-Committee drawn from members of the School Committee. These two bodies would, in turn, co-operate in preparing the educational brief and co-opt other Governors as required. The brief was to be completed in June when it would be discussed by the School and Finance and Policy Committees who would by then be more fully informed of the insurance position and of the inevitable financial constraints.

Meanwhile it was important for those involved to make contact and to survey the ground. Preliminary soundings had already begun. Arup Associates worked in studios combining architects, engineers, quantity surveyors and designers within a single team. This allowed for greater co-operation and speed of working in the design and construction process. But it was a new concept to most Governors and several went up to check on the workings, fees and methods of operation involved. Meanwhile Philip Dowson, Derek Sugden and others in the Arup studio came to meet the

Head Master, Staff Planning Committee and pupils in a series of informal meetings to check out the objectives and general perceptions of the client. Both sides appear to have been reassured by these preliminary contacts. A few even allowed themselves a certain measure of enthusiasm.

The Staff Planning Committee had been appointed ten days after the fire. It met on the first day of the holiday to consider its terms of reference and working approach. Extracts from Bernard Feilden's 1973 Feasibility Study, itself the product of a wide range of staff soundings, had already been distributed. This had previously mentioned the need for more classrooms, subject areas, reading rooms for private study, administrative and social centres as well as the need for audio-visual and computer facilities. It had even commented on the possibility of splitting up the Great Hall area into three levels to provide for the additional space. As the state of the shell became known so thoughts of an entirely new building receded. On the other hand there was unanimous agreement on the need for a Great Hall to act as a focal point and a formative influence in a boy's career. To fulfil its role it would have to be visually 'amazing', visibly accessible, yet acoustically isolated, a fair challenge to any architect! Debate on other areas was exploratory and less decisive. A centralised administrative unit would have its advantages and disadvantages; each department would no doubt have its individual requirements; to build a new combined Junior School, a move suggested in a number of previous feasibility studies, would be both costly and time-consuming – the previous juxtaposition of Upper and Lower School had had advantages for both boys and staff, so the existing pattern was likely to survive. Members were left to reflect during the holidays on their departmental needs, practical and aesthetic concepts, and the School's overall requirements.

Specific investigations began in the first week of the Summer term. Informal meetings with the architects had already brought key areas into focus. Some approximations of the possible floor area had been provided. Departments were each to define their 'ideal' requirements for teaching, study, staff and storage, allowing for as much flexibility as possible. Circulation within the building was to be seen in the context of the whole estate. The functions of the Hall, administration, services and technical requirements were to be carefully specified. Priorities were to be drawn up in case the full list of demands could not be fitted into the space available. The Committee met weekly, hearing or receiving submissions from every user of the building. Surprisingly, departmental patterns were quick to emerge, with large form rooms for Middle and Lower School teaching, smaller rooms for VIth form work and a mosaic of study areas, libraries, staff bases and storage facilities. These could be rapidly quantified into four basic sizes and grouped into a variety of interlocking subject areas ready for the architects to redistribute them within the building.

Plans for an improved Staff Common Room had already been drawn up before the fire and were incorporated with little amendment. Designs for a

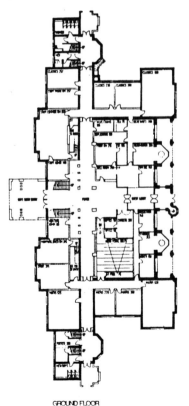

The first outline proposal of the ground floor (Plan B)

GROUND FLOOR

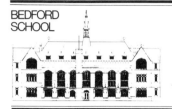

BEDFORD SCHOOL

Development Bulletin

No. 2 March 1980

The second Bedford School Development Bulletin comes at a time when the heart of the School Estate is much affected by builders' lorries, bulldozers, excavators, as well as much appreciated signs of real activity on the Main Buildings. The days of general planning are now behind us and, though there is still much detailed work to be covered, we can now look forward optimistically to the restoration of the Main Building commencing in earnest in time for eventual occupation in September 1981.

The ARUP plans for the Restoration are shown in this bulletin. On the ground floor, the main reception areas and rooms, offices, Common Rooms, and part of the Mathematics Department. On the first floor, the new raised Great Hall is surrounded by the Lower School positioned at the West end and other departmental areas to the East. The second floor is devoted mainly to Modern

Languages and the Lower School at either end, together with galleries for the Great Hall. A new third floor is divided between the History and English Departments.

The North façade remains essentially unaltered, but the South façade will be modified to include a porch entrance leading into the main foyer and through a staircase into the Great Hall.

John Laing Ltd., is carrying out the work under a Management Contract agreement.

The next bulletin will contain more detailed layouts of the areas within the building, the internal furnishings, and items we hope to include in the Great Hall and main rooms. As usual the Ousel will be keeping its readers up to date with developments as they occur on a termly basis.

C. I. M. Jones, Head Master

Model of the New Great Hall.

Photos: Arup Associates

The second development bulletin with photographs of the model of the Great Hall.

new Monitors' Room were submitted by the Head of School. The uses of the Hall and its focal role within the circulation of the building were again emphasised; the importance of heating, lighting and storage facilities were repeatedly stressed. Debate on administrative requirements was resumed. Here the old building had been especially deficient, having been built at a time when administration had hardly begun to impinge upon the educational scene. Now there appeared to be a need for studies, offices, reception areas, interview rooms, ancillary facilities such as cloakrooms, stores and kitchens as well as spaces for both formal and informal meeting and discussion. It was clear that taken as a whole an administrative block could become the dominant feature of the building; the Committee felt strongly that this should be avoided and that the administrative should remain secondary to the educational requirement.

Also excluded were any thoughts of a new Library; it would take up a considerable space and much could be done to embellish the existing Library in the Memorial Building, especially if the Careers Offices were to be moved out and into a new, more central, space in the Main School. Religious Studies might also benefit from separate facilities in one of the surrounding houses, especially when fulfilling a less formal pastoral role. Meanwhile in debate the Committee ranged far and wide, exploring the desirability of carpeting, display areas, ducting and TV and computer cables, telephones, the use of mezzanine galleries as well as a whole range of departmental features such as blackboards, whiteboards, rollerboards and pinboards!

The staff discussions had been joined from time to time by members of the Project Sub-Committee. Minutes had been sent regularly to the Architects as well as to designated members of the Governing body. When, therefore, the Staff Planning Committee's Report appeared at the beginning of June it was possible to orchestrate a quick response. Within two weeks the School Committee had discussed and accepted the Report, forwarded it to the Architects as the Brief for the restoration of the building and met with Arups in joint session to receive the Architects' initial response. The Arup Studio was surpised and impressed by the clarity of the Brief. (They were only to discover later that the Staff had been considering a number of alternative teaching and administrative structures for some years before the fire.) The architects found it 'almost miraculous' that the accommodation requirements, with the exception of Religious Studies and the Bursar's administrative area, more or less fitted into the building. There can be little doubt that the close correlation between the Staff's projected proposals and the possibility of their practical implementation helped to give extra impetus to the re-planning process and to the eventual acceptance of the final design scheme.

From the start the Architects had been influenced by two major factors: the need to provide more accommodation and to relate the building more directly to surrounding School activities. In addition Philip Dowson had at an early stage envisaged raising the Hall into a *piano nobile** at first-floor level, with a grand staircase linked to a new porch at the centre of the rather nondescript south front. These ideas were now shaped into a Hall elevated with spectacular views over the School playing fields and releasing space below which allowed for cross-circulation at ground-floor level, thereby increasing connections to and through the building from all parts of the

Design scheme proposal (Plan C)

---

*The principal floor in a large Italian house, raised one storey above ground level and reached by a grand staircase.

Estate. The new south porch entrance provided direct access between the Main Building and the 'quadrangle' as the Architects insisted on calling the area south of the Main School Building, while the grand staircase provided a spectacular approach to the Hall. In addition the generous roof space permitted the introduction of a third floor which, together with the ground floor arrangement, provided for a 30 per cent increase in the available area.

Before any further plans could be submitted, much detailed work had to be done. The Architects required clarification on a number of unique School features: what were the 'Inky', the Bell Room, the role of monitors, the precise functions of the Great Hall? The arrival of the first drawings at the beginning of July opened up a lengthy debate on the planning of the ground floor. The Arup Studio had always felt that the staff and administrative accommodation should fit neatly under the Great Hall. In examining the files four years later Michael Lowe, the architect involved, discovered that Arups sketched the ground-floor layout five times before settling on a preferred solution. They also found a further five sets of suggestions from the staff on how best to answer their requirements. His account gives a good indication of the way the Architects worked at one area of the building and on their relationship with the Staff Planning Committee:

The first plan A, which was rapidly drawn freehand, suggested that the Head Master (and Conference Room), together with the Staff Common Room be located on the south side flanking the new entrance. The north entrance, which was always accepted by everyone as the formal visitors entrance, contained a reception desk rather like an hotel! The common facilities such as the audio-visual room, careers, duplicating and storage were consistently placed centrally under the Great Hall. In Plan B which represented the first outline proposal, we buried the Head Master between the Bell Room and his secretaries. The staff and boys had no direct access to him. We also finally understood the function of the Bell Room and managed to locate it adjacent to the north entrance so that it could double as a School 'base' and reception room. The Staff Common Room and Monitors were now placed adjacent to the south entrance and were able to survey the quadrangle. This seemed to satisfy everyone and some progress was being made. Finally in Plan C we reluctantly bowed to pressure from the staff to relinquish some handsome west facing rooms that we had reserved for the Classics Department, and located the Head Master in the north-west corner of the building. His secretaries and the Usher completed the plan for that zone of the building. The new layout provided a passage link from the Staff Common Room to the Head Master's Study. The boys also had easy access to the Head Master. In addition the Head Master was in a good position to view activities on the northern side of the School, including the front entrance gates.

However, the process was to continue with Plan D which formed a

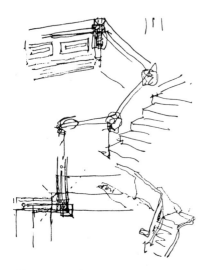

The new staircase

All line drawings by Julian Bicknell.

Great Hall capitals

part of the School Design proposals. We were still juggling with the arrangement of the audio-visual room, Careers and various interview rooms. We also re-arranged the structural columns in the Lower Foyer and the main staircase flights up to the Hall. Finally in Plan E, when we had practically run out of time, we settled on an arrangement for all the central common facilities with the addition of a staff coats room that appeared late in the day. The columns had moved again and we, at last, managed to resolve fitting in the staircase with the help of a waiver from the North Bedfordshire Borough Council. The Head Master's study had mutated and become subdivided to provide him with a study and adjoining workroom. Shortly afterwards the builders moved on site and we could make no further changes.

Luckily, other areas did not present quite as many problems.

The third floor planning proceeded rather well. We recognised that we could make only minimal changes in the external appearance of the Grade II listed building and managed to devise a system of rooflights that were set behind the existing ridge line. This allowed us to create a series of well-lit attic classrooms. We were very keen to present the feeling that you were located 'in the roof'. However, we did manage to include some small windows which provided spectacular views over the School grounds and the town beyond.

Departmental areas were slotted into appropriate parts of the building with English and History sharing the new top floor, Modern Languages occupying much of the second and Mathematics the ground and first-floor areas at the east end and adjacent to the Computer Room. The Lower School returned to improved accommodation on the first and second floors at the west end, Classics, after several moves, being eventually settled in a prestigious position on the first floor at the top of the grand staircase.

While considering the details and adjustments contained in the various Arup proposals and sending back their own comments and drawings, the Staff Planning Committee was also examining a wide variety of detail from blackout facilities to cleaners' stores. Besides studying the layout of the main services such as lighting and heating they were to cover with the architects the entire furnishing and fitting out of the building from floor finishes to electrical fixtures. It would be tedious to mention all the details even if some, like the size of common-room pigeon-holes and the ramifications of a new telephone system, did provide an element of entertainment. It was hardly surprising that several members began to feel that they had indeed taken on an additional careeer.

Meanwhile the Governors were receiving and assessing a series of architectural presentations as part of the crucial planning and decision-making process. The first of significance was on 17 July when Arups brought along their initial Outline Proposals first to the Joint Planning Sub-

Committee, which combined Staff representation and Governors, and then to the School Committee. Arup proposals were always attractively set out with booklets in dark transparent perspex covers that helped to heighten the clients' expectations. The contents indicated that they could fulfil 95 per cent of the Brief but at a cost rather higher than had been anticipated. The plans appeared imaginative, the elevated Great Hall impressive and the use of space allocated to Departments practical and effective. For the rest of the coming year the Governors were to be faced with the problem of how to finance a scheme which they clearly wished to implement but which was going to be costly and was in danger of gaining a momentum of its own, and one that could become increasingly difficult to contain. At this first critical meeting they agreed that, notwithstanding the cost, they should consider implementing the whole scheme as one 'worthy of the School' and as an 'act of faith' in its future. This recommendation was confirmed a week later by the Finance and Policy Committee in spite of continued anxiety about what was seen as an inevitable shortfall between the monies to be received from insurance and the eventual cost of a modernised and enlarged 'new building'. It was felt that an additional £1.3 million might possibly be raised, but not much more, so the inevitable sub-committee was set up to look at the residual difference and consider ways by which it might be covered.

During the next two months the Governors, Trust Officers and Head Master were all considerably exercised by this financial problem. Two meetings with the Architects brought a number of modifications to the Outline Proposals; these virtually bridged the gap between the cost of the whole undertaking and what the Finance Committee felt the School could just about afford to pay. A whole series of memoranda drafted by the Finance Officer, the Chairman of the School Committee and the Head Master gave a variety of possible sources from which to balance the expected deficit. There was general agreement on the need for a fresh Appeal and the possible use of part of the endowment. The Governors were from the start anxious to avoid any additional charge on the fees, and the Chairman had stressed their special responsibility in covering the financial side of the whole operation. But some debate appears to have occurred between those who felt that the additional improvements ought to be self financed and, therefore, come largely from an increase in the fees, and others who thought that the Harpur Trust as the Governing Body should use up some of its capital resources to make a significant contribution.

Sketches for the Great Hall roof.

These issues were still undecided when the Governors came to consider the Architects' final Design Scheme on 26 September. Luckily the insurance claim had by now been settled.

The building had been insured for £2.15 million. After negotiating the value of those parts of the structure remaining and other issues under the terms of the policies, the Governors opted for an immediate indemnity payment of around £1.9 million which, invested over two years, would, with high interest rates, bring in another £300,000. The atmosphere of the Arup

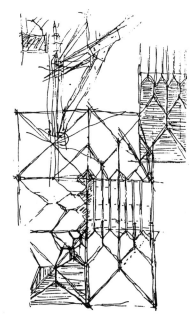

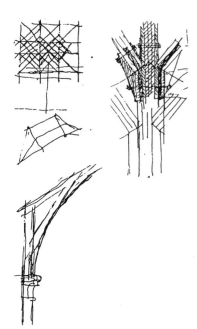

presentation, the well-considered and accurate detail in the proposals and the knowledge that a significant proportion of the sum involved was already in hand brought from the School Committee a unanimous vote in favour of the full Design Scheme. This was immediately followed by an endorsement from the full Board of Governors. At the same time John Laing Construction were accepted as the management contractors. The cost limit was set at approximately £3.5 million and this was to remain the target figure for the entire operation. Construction costs as at June 1979 were £2,125,000 with a further £425,000 allowance for inflation. V.A.T. at 15% and comprehensive Design Fees at 17% together with the cost of site supervision and safety measures brought the figure to a formidable £3,462,849. The various methods of meeting the shortfall would now have to be vigorously pursued.

The restoration of the School had now been effectively launched. For the next three months Governors, Staff and Architects were generally to go their

The assembly of the roof is complete. The panels are constructed of cedar and they rest on timbers of Russian redwood, themselves supported by corbels of steel and concrete. A point of interest is that the Arup design provides effective support for the extensive roof without any cross-ties, a remarkable achievement considering the pressures that can be exerted at this height by gale-force winds. It makes an interesting comparison with the structure of the original roof. *(John Laing plc)*

separate ways and concentrate on different areas, not that there was any lack of communication between them, for this was to be a uniquely friendly and harmonious enterprise. But while the Governors remained occupied with finance and policy, the Architects had to complete their Great Hall designs and hurry through the working drawings in time for the builders to start operation in the new year. At the same time the Staff Planning Committee, besides checking the Architects' plans, had to concentrate on furnishing and fitting out the building while simultaneously attempting to teach and carry out their other school obligations.

When the parents' elected governor arranged briefings to explain the Design Scheme and the costs involved, renewed discussion was stirred as to whether the financial deficit was to be covered by fee increases or by a contribution from the Trust's endowment. The general reticence of any finance committees to publicise their procedures led to some anxious questioning. When those involved widened their pleas for assistance to include the needs of the other three Trust Schools this seems to have gone down rather better than earlier attempts at special pleading! In the end agreement was reached that there would be a further reappraisal of the Trust's assets. A meeting of the Finance and Policy Committee early in the new year was to set the machinery in motion. By midsummer 1980 it was revealed that the Trust would be able to finance a loan of £700,000 towards the restoration of the School from increases in the endowment revenue. The financial target had come within reach!

Meanwhile the Staff had become increasingly enmeshed in drawings and detail. The Architects came to interview all the main users of the building individually. The plans of every department and user zone were checked and double-checked. Lists of furniture and fixtures were catalogued and their position finalised. Some features were repeatedly questioned: the height of coat hooks placed on a dado one metre above the floor; architects no doubt wore donkey jackets but the School Rules specified knee length coats or mackintoshes! In some areas the staff took a decisive initiative: in urging the need for soft floor coverings; in rejecting the idea of an intercom system from the Head Master's Study to every room – was the Committee perhaps overconscious of the approach of 1984? New areas of investigation were always opening up: the need to study alternative blackout blinds, organ designs or furniture construction. The Committee had held ten meetings during the Christmas term bringing its total to twenty in the course of the year, twenty-one if one included its preliminary meeting. It had certainly come of age!

The Architects' presentation of the Great Hall Design plans to Staff and Governors on 18 and 19 December was to bring the year to a fitting climax. Once again there were the carefully printed and presented booklet, slides and documentation, but this time with a model so constructed as to provide a mirror image which appeared to the viewer to open up the full dimensions of the new Hall. This, if anything still needed to do so, was to confirm in

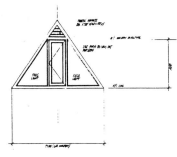

Sketches for the new top floor.

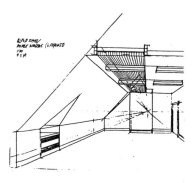

All line drawings by Arup Studio.

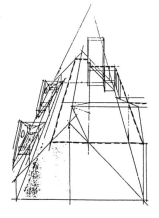

(Chris Brown)

Above. Symbol of renewal: the vine flourishes. *(Arup Associates)*

Left. After the inferno. *(Arup Associates)*

Above. Aerial photograph by Mr Douglas Butler showing the crane which conveyed building materials into the School. *(Douglas Butler from a helicopter piloted by Charles Lousada)*

Right. This picture helps to explain why the demolition firm took from the middle of March to September to prepare the site for rebuilding. *(Arup Associates)*

Left. Classroom village. *(M. P. Stambach)*

Below. Phillpotts Memorial Gates. *(Morley Smith)*

Right. New South Porch with conference room window above. *(Morley Smith)*

Left. General view of the theatre. *(Morley Smith)*

Below. Recreation centre – general view of the Theatre and Sports Hall. *(Morley Smith)*

Right. An interior view of the refurbished library. *(Morley Smith)*

Overleaf. The Gala concert October 1981. *(Morley Smith)*

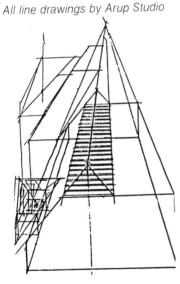

everyone's mind the conviction that the efforts and commitments undertaken so far would indeed be ultimately worthwhile. Skilfully lit, the whole was a veritable harmony in wood and stone. A floor, carefully stepped and graded, for use both as an amphitheatre and for a more traditional layout, provided a base from which pillars, galleries and tracery windows rose to a tapered roof latticed with a pattern of closely supporting beams. The Architect's sketches show how the design evolved from a brief flirtation with the hammers of the old Great Hall through a variety of geometric modifications to the final whole. A previous roof designed by Arups for concert hall use had been at the Maltings, Snape. Here, a boarded timbered roof was supported by four trusses fabricated from timber with steel tie rods. The remnants of this idea can be clearly seen in the drawings of the Hall submitted in the September Design Scheme. In the new Hall, however, the Architects wished to achieve a sense of upward thrust and greater height both to provide the necessary acoustic space and to catch the eye in a dramatic perspective. Julian Bicknell and Charles Wymer had made a model of spaghetti in order to experiment with various alternative designs. Unfortunately it was never photographed! As the roof rose more steeply so the side struts inevitably fell away. As the height increased so a pattern of

All line drawings by Arup Studio

Right. A top floor classroom.
(John Laing plc)

branching tree-like supports grew up and separated out into a series of diamond-shaped diagonals. When the engineers hit upon the idea of laminated beams rising from the capitals and subdividing into ever finer tracery patterns the whole emerged complete. The presentation gave a fair impression, but the model provided a much more accurate rendering of the final effect as the many photographs taken inside it clearly show. The impression was overwhelming.

With the presentation of the Great Hall the initial planning phase came to a close. The future was to be a matter of more and more detail as working drawings poured in an apparently endless stream from the Architects' offices. The insertion of the Great Hall plans into the Design Scheme at a later and a separate stage was to create some future difficulties. A number of issues such as the storage of staging, spot lighting and the servicing of parts of the ventilation system became telescoped and were never fully defined. But the pressure of time was unrelenting and was to allow for minimal revision. In the new year building had to begin and Laing's offices were first erected at the north-east corner of the site area and later moved behind the Dining Halls when external works were progressing. It had been impressed upon the planners that any modifications were bound to have cost implications. In any case many would have admitted that their creative and critical energy had become virtually exhausted! A friendly Clerk of Works had been appointed to look after the School's interests and the Staff watched with relief as the building, now cleared of rubble, was prepared for its impending reconstruction. Only later, nearer the time of the handover, were there moments when some had their doubts! But those days were still a long way off. At the end of the year Governors, Planners, Officers and Architects could all feel that they had come through with flying colours and look forward to a worthy conclusion to all their labours.

Chapter V

Reconstruction

1980 was to be the year of reconstruction; the School had to be rebuilt by the end of the year if it was to be ready for reoccupation in the following September. It was to be a time of hard work and occasional anxiety for Governors, Staff planners and the Management team. Ideas had to be implemented in detail and this inevitably created new difficulties. Patience and perseverance were called for, especially by those unfamiliar with this sort of enterprise. Gradually the building was seen to emerge from within its shell, first in outline then in substantial shape. But progress on a reconstruction project such as this was very dependent upon the stability of the existing structure. Many factors could not be assessed until a proper examination of the walls and foundations had been made and when various areas of the building had become exposed to view as a result of demolition and early construction activities. This inevitably meant modifications to the design and construction methods and in some cases, of course, increases in cost. There were also the problems of financing the short-fall, of VAT, and of launching the new Restoration Appeal. Through it all the teams worked doggedly on. There was no alternative.

The Staff Planning Committee was not disbanded early in 1980 as had originally been anticipated. It was to hold another forty meetings – the last in December 1981. Besides dealing with the furniture, fixtures and fittings, it was to handle anything from Great Hall chairs to the latest telecommunications technology. It was to consider landscaping and car parks, the re-allocation of space freed after the reoccupation of the Main School Building, and the positioning of departments which had not been included in the plans because of lack of space. Throughout the year the Committee held a monthly joint meeting with the Architects at which all the details were examined, plans inspected and alternatives discussed. The Governors followed events with a parallel set of joint meetings at which the Architects reported the latest financial and structural position, brought up areas of difficulty and reviewed any staff proposals. They also set up and assessed the

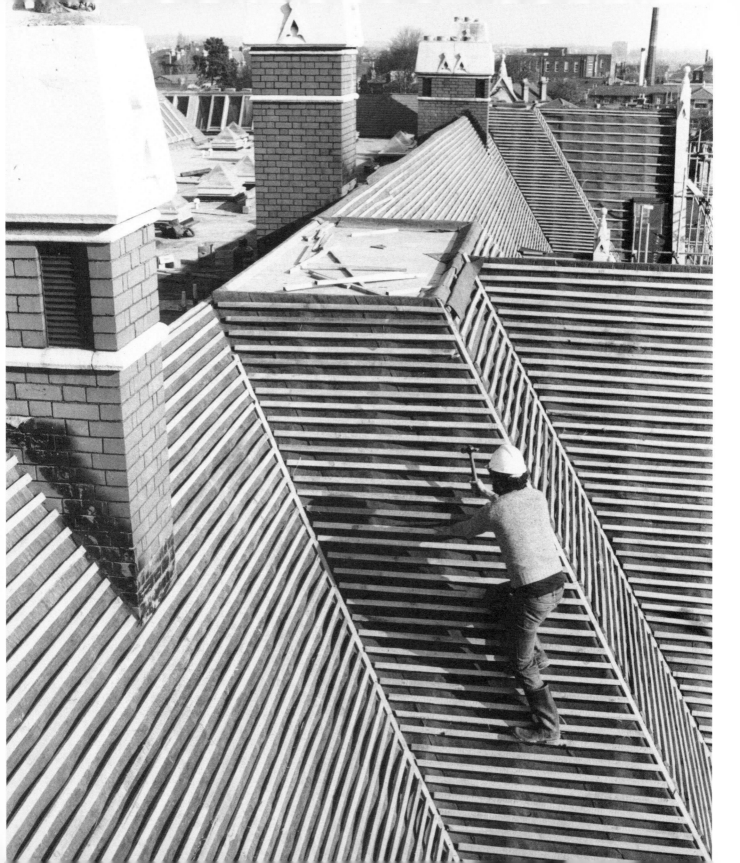

Eleven miles of battens on the roof. *(Times Newspapers plc)*

The building nears completion. In the northward projection of the roof on the right of the building the windows lighting the History Department can be seen. The reclaimed attic area of the roof houses the English as well as the History Departments. Included in the restoration work was the cleaning of the entire exterior of the building in order to remove the traces of the fire - and of age - in order to match up the colour of the old stone and brick with the new. Here the contrast between the two is evident. At this stage the cloisters have not yet been filled in to provide form rooms and offices. *(John Laing plc)*

work of the Restoration Finance Committee which co-ordinated the Appeal, and followed up the question of VAT, while covering all the routine work involved in the general running of the School.

Completing the structure was clearly the first priority. Here control over the programme was a critical factor and meetings of the School Committee held jointly with the Architects were to occur every two months to hear reports of progress and compare tender figures with the Architects' original estimates. The first return which was £21,000 above target caused a slight tremor, but the second was below, the third in balance and the fourth £50,000 below. These ups and downs were to alternate throughout the year. The programme, on the other hand, seemed to drift into arrears. By April, after a period of bad weather, it was reported half a week behind; in June it was two weeks late, owing to a shortage of bricklayers, by September three and a half weeks because of difficulties with the roof. Laing's restructured the programme in the late autumn in order to make up for lost time and assured the Committee that the target for the following July would still be met. But, owing to unforeseen circumstances, the drift tended to reappear. In May it was again four weeks behind and completion was extended to the

end of August, allowing only two weeks to set things up for the beginning of term.

But the work was done, and Governors and Staff could wander round and watch as plans acquired new dimensions. The Hall roof could be seen in its skeletal symmetry, the tracery windows with their delicate carving, the staircase in its simple yet imposing design. Carpenters and stonemasons could be viewed at work. The quarry at Weldon from which the School's stone facings and decorations had originally come had recently been re-opened, so the new stonework would eventually blend in with the old. But in the raw there was something both expectant and awesome in the vastness of an area as yet untouched by those finer finishes that make buildings so comfortably reassuring and obviously secure. The patchwork of old and new, brick and concrete, stone and cement, laced with a maze of leads and ducts for the electrical and electronic services, stretched out like some great carcass not yet come to life.

C SECTION LOOKING NORTH

Sketch of the organ gallery
(Arup studio)

The Head Master, Vice Master, and the clerk of works surveying the Great Hall. *(Times Newspapers plc)*

Worth noting in this picture are: the concrete reach of the floor, raised now to first floor level, scored with the electric power lines feeding outlets to the ground floor rooms; the brick facings of the concrete pillars supporting the stone arches. These arches replaced the old stone ones which were virtually destroyed; the stone columns climbing towards their clearly visible destinations. In fact the latticing of the roof is supported by steel corbels integrated into the concrete pillars; the lateral concrete span on which the east gallery is to be based.
(John Laing plc)

Meanwhile, morale was sustained by the special occasions, the unveiling of the Commemorative Stone and the Topping Out ceremony, by the arrival of new men with sophisticated crafts and skills to build the organ, carve inscriptions or install new telephones, and especially by the encouragement and support invariably given by all those with whom contact was made in the course of the reconstruction. The County Supplies Office was to provide an opening to a wide range of firms and manufacturers. Avon Cosmetics made special provision to allow members of the Staff Planning Committee to inspect and test out their epoxy painted corridors. Churches opened their doors to allow others to hear and play their organs. The reconstruction struck many a sympathetic chord, and gestures of encouragement and good-will did much to sustain the effort and enthusiasm of those bent over plans or poring over furniture and lighting schedules. For them the work had become less creative, more routine. Everything had to be checked and re-checked, or rather as much of it as possible. Nagging doubts raised had to be

resolved; most were accounted for but a few were to slip by. The building was to bear witness both to the general success and to the occasional failure.

It would be tedious to enumerate all the detailed work of those two years, but a number of highlights do remain either to instruct or to entertain. Furnishing the building was to be a major undertaking. Every form room required up to a dozen items from large and small tables, master's desks to waste bins. Study areas and departmental libraries had to have worktops, cupboards, shelving and chairs. Then there were the offices, Staff common rooms, waiting areas. The Bursar and members of the Planning Committee visited schools, collected catalogues, viewed the manufacutre of work tops in Stevenage, curtain materials in County Hall. The Usher produced endless lists specifying the requirements of every room – there were over eighty! The Bursar interpreted the views of the Staff planners by ordering items from over half a dozen suppliers. To beat inflation and prepare for the final move back into the building as much as possible was ordered well in advance. This was gradually set out in the temporary classrooms or stored wherever there was spare cover to be found on the Estate. Everything was to be ready for the day of the reoccupation. But would it all fit? Well, it did!

Blinds were to be one of the minor bones of contention. These were needed for blackout or to shade the skylights on the third floor. The staff tried repeatedly to emphasise that blinds were fragile things and, therefore, at risk in any school environment. Unfortunately the sort of mechanical controls which had existed in the past were either too expensive or unavailable so choice was limited to a screw mechanism or nylon cords. The Clerk of Works attempted to demonstrate the strength of nylon cords by abseiling down a wall with near fatal consequences; in spite of this clear warning nylon cords or screw handles it had to be! In due course the Staff's anxieties were to be fully confirmed.

The Great Hall chairs, on the other hand, were the successful outcome of one and a half years of combined research and experiment by the Architects and the Staff Planning Committee. To find chairs appropriate for the Hall as conceptualised by the Architects was difficult enough; to find ones which fitted the requirements of a school as well as Fire, Borough, Planning and EEC regulations proved virtually impossible! The Head Master laid down that the chairs were to be strong, stackable, comfortable, linked, appropriate to the design and within the cost limit. But such chairs clearly did not exist. Slowly at first but then in growing numbers chairs from over half a dozen suppliers began to accumulate in hallways, waiting rooms and other vantage points. Some were too large, others would not link easily, many were too expensive. The Architects favoured flowing curves or contemporary shapes; the staff were a little more cautious and conservative. 1981 came and still no solution had been found. An order would soon have to be placed or there would be no chairs. Eventually the User thought he had seen a possible model at ESA. It was acquired and examined. The manufacturer agreed to a number of modifications to fit the School's needs. The

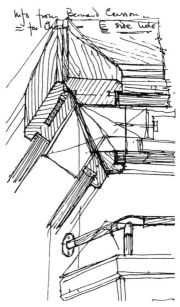

The building rises within its shell. *(M. J. Norris)*

The Great Hall gallery. *(Julian Bicknell)*

Architects took the chair away to study it and to assess its likely impact on the overall design. In the end it was approved and ordered, specially modified to blend into the Hall and harmonise with the woodwork and the panelling. It did not quite fit every back or help lull the weary during long concert programmes, but it did assist the concentration and provide the necessary support.

The purchase of an organ appears to have caused less difficulty. The old Hall had had a nondescript electronic organ to accompany morning assemblies; its loss had been one of the few unmourned casualties of the fire. It was clear that an organ would be needed and that every effort should be made to acquire one which would be worthy of the new building. The Staff planners felt that as the instrument would have to be incorporated into the design it ought to have priority over other items of furniture at the planning stage. The Director of Music and Chapel Organist were already aware of an organ builder who would be able to supply a suitable electronic instrument built to pipe organ specifications and Copeman Hart had been approached in 1979 with this in mind. The Planning Committee listened to tapes which appeared to reveal no noticeable difference between these new electronic organs and traditional pipe instruments. In April 1980 members joined up with the Architects and their acoustics expert for a tour of Copeman Hart instruments in Crawley and Folkestone. As a result a list of alternative specifications and costings was drawn up. Clearly all hoped to gain the maximum benefit at a reasonable price. After some discussion the School ordered a four-manual instrument with three manuals complete and the fourth prepared for. It would be finished when the funds became available. In the event, thanks to the Restoration Ball, all four manuals were to be completed and voiced during the course of the following year.

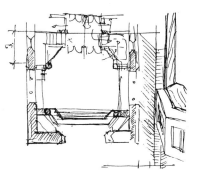

South Porch, ground floor entrance. *(Julian Bicknell)*

To set off the loudspeakers and provide the necessary casing the Architects designed an organ screen which was to become one of the features of the Hall. Set at Gallery level, it blended elements of the traditional design with the new Hall panelling, embellishing the Tower alcove and adding a focal median immediately behind the Head Master's platform. The towers for the 32-foot stops were built unobtrusively into the side walls while the swell was tucked into a back corner. The sounds which emerged were of amazing variety, reflecting the ability of the performers with obvious accuracy. While some felt that a pipe organ would have been more appropriate, lack of space as well as cost made it impracticable. In the event all felt well satisfied with an instrument which was to arouse considerable local and national interest.

Set upon the platform in front of the organ screen were to be the table, lectern and chairs presented to the School by the Mothers' Guild. Their design prompted new dimensions of aesthetic debate. The intention was that they should blend with the screen, but the first drawings and model of the table featured a copper wire grid between trestle and top, no doubt to conceal the speaker's feet, which reflected the loudspeaker grills at Gallery level. This the Staff planners found rather too contemporary, so the table was modified to relate to the screen panels instead. The chairs covered in pale calf leather were also part of the ensemble. Some would have preferred a more striking contrast, but members were becoming accustomed to the Architects' subtle colour sense and accepted their seamless textures as part of the overall scheme.

Sketches of the organ. *(Arup Studio)*

Subtlety of colour had been a feature of many of the model patterns presented by the Architects for the administrative and circulation areas. Shades of beige or off-white appeared to ascend in layers from floor to ceiling like so many levels of sea fog becalmed in the morning light. The Staff were worried that normal paint below dado level would not stand up to the wear and tear that might be expected and suggested that some reinforced surfaces be used. Cost was once again to be the main problem, but in the end an epoxy resinous paint was found and the champagne colour helped to brighten up the passages. Elsewhere in the building the contrasts were to be a little more robust. It was decided that each floor should have two different tones for the north and south facing rooms and that each department should have a distinctive set of coloured pin boards. A Colour Committee including members of the Art Department and the Head Master's wife was set up to harmonise the various alternatives and a wide spectrum of combinations was chosen for the many teaching rooms throughout the building. This, when set up against the plain waxed wooden door frames and dado rails, gave the whole a certain style which appeared to blend contemporary and classical features in a unique manner. It was to be another of the Architects' distinctive contributions.

A special feature was to be the Conference Room situated within the roof space of the new South Porch. During preliminary discussion some had been sceptical of the need for such a room and its name had been changed in the Design Scheme to suggest an alternative use such as reading. Whatever its title it was eventually to emerge as a splendidly panelled room hung above the South Porch with a wide bow window looking out over the southern 'quadrangle'. It even had a secret fire stair carefully concealed behind the panelling. Rather than purchase new furniture for the room it was allocated the two large Thompson mouse tables which had been rescued from the fire and a set of donated chairs which had previously been little used in the Memorial Hall. These, though perhaps rather large and a little incongruous, were to create another more intimate and unconventional design feature.

Conference room above the South Porch. *(Morley Smith)*

Great Hall with organ screen and platform furniture. *(Morley Smith)*

Another difficulty was to deal with the historic furnishings which had been lost or salvaged from the old building: portraits, shields and inscriptions recording great men or significant occasions. It was difficult to assess what proportion of the insurance moneys should be used to replace these losses. Clearly no amount of cash could in fact make up for their destruction and a number have not yet been replaced. The portraits of Head Masters going back some 150 years were of special importance and it was decided to restore these by copies made from the many reproductions available. Fortunately there was an original of the great J. S. Phillpotts in the hands of the Phillpotts family and its measurements were used as the model for the other Head Masters. The whole operation was taken in hand by the Old Bedfordians Club, with the portraits designed to be hung on the south wall of the Great Hall colonnade. Three shields had been saved from the fire as they were away being cleaned. These emerged to help decorate the new staircase. An inscription which had summarised in Latin the early history of the School had been a prominent feature at the foot of the first-floor Gallery in the old Hall. This was now carved in wood and set below the line of balconies in the new. Mr Grasby, who brought with him a fresh appreciation of our efforts, was also to carve the stone plaques commemorating the quatercentenary visit of Princess Margaret to the old Hall as well as the official reopening of the new one, which were to be placed on either side of the new stairway.

As well as with aesthetic matters the Staff Planning Committee was involved with the whole question of telecommunications and the impact of the new technology. With an extra floor and additional administrative space it was clearly important that internal communications should be improved. Initially it was assumed that a manned telephone exchange would be part of the Bell Room, but further consideration amended the plan to cover the whole Estate and raised the issue of what happened in the evenings after school when the central offices closed down but activities in the new Recreation Centre and elsewhere were bound to continue. Discussion, therefore, concentrated upon a whole range of new automatic systems which the Post Office were developing regionally or nationally at this time. The critical issue was whether the Committee and local Post Office representatives could between them find a system which would be both suitable and operational in time for the opening of the building. In the course of the year the Staff were to investigate in varying degrees of depth some half a dozen systems. Some were exceptionally elusive and having been introduced were suddenly whisked off the market. Others remained experimental or only available in remote areas of the British Isles. Eventually the School settled on an entirely new Herald call-connect system which had just been developed. It consisted of eleven in-coming lines each with a mini exchange allowing contact with up to thirty-six interconnected extensions. The installation caused a certain amount of amusement and anxiety as it was clear that the system was as unfamiliar to the engineers as to the users. But

Great Hall with south side galleries and main entrance doors. *(Morley Smith)*

after some initial teething troubles it worked well and it was gratifying to have been able to pioneer something new in this field. With audio-visual and computer apparatus the School was on firmer ground, thanks to the strength of its own departments. Here the problem was to ensure that the desired plans were effectively implemented. The ducting was only to be tested after the building had been opened. It was then rather late to discover that some channels were sharply angled or appeared to be filled with cement!

Outside the building, landscaping, car parking and the general flow of traffic around the Estate also required consideration. Special provision for this area had been deleted from the original Arup contract in order to save money. It was now to be reconsidered as part of the overall restoration of the School Estate. A number of separate problems had to be examined and co-ordinated: the area immediately around the Main School Building, still very much the Architects' concern; the area around the newly opened Recreation Centre; the site of the temporary classrooms after they had been removed; the need for car parking for the new Theatre and for the Great Hall. The Trust's Surveyor was allotted the task of co-ordinating all these sectors and combining the ideas of the Architects, the Staff and the Governing Body. A series of feasibility studies and discussions occupied School Committee and Staff Planning Committee meetings. Consideration had to be given to access by Staff, boys, drama companies, visitors, team coaches and fire engines. The Architects were anxious to surround the building with grass and trees. Flow charts poured out in profusion before the final principles were defined. Traffic would be kept to the west side of the Estate with parking on the periphery, except on special occasions. The east side was left to pedestrians, with no through traffic crossing the north front of the Main Building. Here the grass was to come right up to the Cloisters, showing them off with a new prominence, while a narrow pathway allowed access to the North Porch and across the area for pedestrians, occasional traffic and the usual fire appliances.

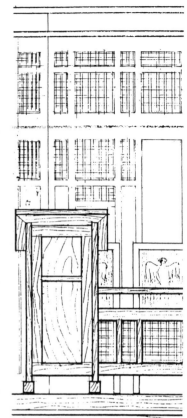

First platform furniture designs. *(Arup studio)*

These additional features inevitably led to a gradual but steady escalation in cost. The Governors kept anxious watch as extras mounted, but while the tenders came in below estimate some degree of flexibility remained. In October 1980 the Finance Restoration Committee launched their new Appeal with a brochure which appeared suitably tinged with flame and ash! It aimed to raise the extra million pounds needed to cover the contract and the additional costs, including the £422,000 estimated for VAT. It was to be divided between the local Appeal Organiser, who had done so well in raising money for the Recreation Centre, and the City, where it was hoped to interest a number of finance houses and charitable institutions. But while the local Appeal got off to a good start the City was slow to respond at a time of recession and high interest rates, a fact which was to make the VAT issue of increasing importance.

The question of VAT had been delegated to Mr G. C. W. Beazley, Vice-

Chairman of the School Committee. For two years he had burrowed mole-like into the law, emerging from time to time with statements of Delphic discretion. MPs had been lobbied, Counsel questioned and precedent explored. In essence the question was whether a building so entirely re-shaped and restored was classified as a repair or as a new work of reconstruction. In the former case tax was charged at 15 per cent, in the latter it was zero rated. A number of test cases were proceeding through the Courts at this time and one had on appeal reached the House of Lords. This appeared to allow the division of the work between what was new and what had been repaired and Architects and Contractors had by the end of the contract period divided it up, reducing the tax by about a half. Another test case, however, was going through the Courts, that of a church, St Luke's, Crosby, which had also been burnt down and rebuilt within its shell to a new design with reinforced foundations. But this would not be heard until 1982; so everyone had, in the meantime, to remain in suspense.

Just before the hand-over and shortly after prescient warnings by the Chairman of the School Committee, the suspense was heightened by the only real hiccough of the entire operation. Early anxieties had after all been more than justified; delays had, at their worst moment, held up progress by up to ten weeks, though this had been overcome by additional effort on the part of the Laing management team. The hold-up and other misunderstandings had, however, left a certain number of claims and counter-claims between various sub-contractors, as well as a need to pay for the month's extension of the contract. As a result the apparent surplus on the tender figures suddenly evaporated, as did the construction reserve contingency which had remained largely untouched until that moment. Pointed questions and critical comments were voiced for the first time. Additional items would not have been incorporated into the building had the Governors known sooner! Cuts could have been made! But all was to work out well in the end. The Architects were able to negotiate significant reductions in the claims, many of which cancelled out. In the case of St Luke's, Crosby, the judgment of Mr Justice Woolf allowed for the possibility of a work of reconstruction incorporating part of an old building, including its walls. The Commissioners of Customs and Excise did not appeal. The reconstructed building was to be zero rated.*

As a result of the VAT verdict the final financial equation was to come out rather better than might have been anticipated. The total cost came, if by a somewhat roundabout route, to the very £3.5 million figure originally agreed. Construction costs had increased by some £175,000 since the Design Scheme presentation due to rising inflation and the difficulties associated with the completion of the contract. The addition of the landscaping, new fittings, telephones and Great Hall staging had, with other elements, further inflated the total to around £3.9 million when the VAT judgement

*It is worth recording that the Governors made a presentation to the church in recognition of its contribution in pursuing its case to the High Court.

intervened to reduce the final figure to some £3,502,942. Of this, £3 million was covered by insurance moneys, including interest, and a bank loan financed from the endowment, leaving £500,000 to be raised from the Appeal. Even allowing for any extra costs for the additional car parks and the levelling and re-seeding of the temporary classroom area, the excess was negligible. The classrooms themselves were exactly covered by the insurance and by their subsequent sale – the last five units going to another school hit by fire! On 28 August 1981 the restored building was handed over, a hundred weeks after the Governors had accepted the Architects' final Design Scheme – four weeks late! It was a near record achievement.

A year was to be spent in dealing with the snagging, the loose ends and un-finished items. To some this came as rather a shock, though subsequent experience has brought a greater sense of realism. One notable event was testing the acoustics of the Great Hall. Derek Sugden arrived from Arups equipped with pistol and measuring gear. The Hall was emptied of chairs,

Third floor classroom 'in the roof'. *(Morley Smith)*

then filled up again. Some seventy-five shots were fired to check the sound decay at a representative range of frequencies. In the end he declared himself satisfied. Subsequent concerts have fully confirmed his confident predictions.

The School reoccupied the building with a rush, like so many lemmings emerging after years of darkness into the light. Fixtures were soon tested to breaking point and those not up to standard quickly cracked under the strain. Plastic door stoppers were no match for heavy chestnut doors swinging hundreds of times a day! Handrails with one-inch screws could not be expected to take the strain of a thousand schoolboys' hands. And as for tamperproof thermostats, the Staff Planning Committee's worst fears were all too quickly confirmed. But the Hall was to create its own magnetism, the classrooms their own individuality, and the whole layout its daily flow of human activity allowing each area to function as a separate entity within an integrated framework, while the artistry and excellence of the concept and its design would make a lasting impression on all who viewed it then and in the future.

Thoughts of gratitude were in every mind when the Chairman of Governors gave a dinner for Planners, Architects, Management Team and Insurers at a Glazier's Hall towards the end of that year. As each participant surveyed his presentation goblet, with an eagle engraved on one side and a phoenix on the other, he could reflect on the experience of two and a half years. Of plans conceived, drawn, amended and completed; of contingencies and crises covered and catered for; of hopes realised; of minor disappointments; of detail to be corrected; and especially of the achievement of a unique and combined effort involving many teams in one carefully dovetailed and co-ordinated enterprise. All could feel a justifiable pride at having been associated with so successful a project.

With thoughts of the present and of the past were mingled thoughts of the future. Schools are more than their buildings, more than their pupils, more than those who work the Prospectus; they are a combination of all of these. But the buildings divide the space and create the atmosphere to help those who teach and those who work towards an attainment of their goals and a definition of their values. The new building had been designed by architects sensitive to these things. Would the future fulfil their hopes? Only time could tell how far expectations would be realised; whether the building would leave the same imprint on future generations that the old had left upon the past. But for those who had been through it all there could be no doubts. It was an experience none would have wished to miss and few, if any, would have wished repeated.

Chapter VI

The New School

When the Main School Building was reoccupied in 1981 the general mood was one of exhilaration. Few were concerned about the snags or minor defects which could safely be left to sub-contractors or occasional committees, (though boys being critical by nature did not allow the odd blemish to pass unnoticed!). The clean lines, brighter tones and new perspectives all invited enthusiastic comment and showed up rather less favourably some of the more jaded sectors of the School Estate. All could see that the new interior stood out in style and structure when compared to the old.

The southern 'quadrangle'.
(Morley Smith)

Ground floor corridor, 'clean lines, brighter tones and new perspectives'. *(Morley Smith)*

Re-designed science labs. *(Morley Smith)*

The Lucas wing. *(Morley Smith)*

The refurbished Memorial Hall. *(Morley Smith)*

View of the library foyer from the Memorial Hall staircase. *(Morley Smith)*

Risen by John Tavener *(Chester Music)*

The events of 'Restoration Year' displayed the Great Hall and its qualities to full effect. At the Service of Dedication the Bishop of St Albans, while emphasising that the spirit of a school is more important than the buildings, nevertheless added that 'to be surrounded by things of beauty, by fine architecture, by excellence of any sort, can have a deep impact on the lives of all who learn here'. The opening concert with its resounding choruses of *Risen* by John Tavener, especially commissioned for the occasion, the organ recitals led off by Sir George Thalben-Ball, the jazz performances and official functions all revealed its many variants of layout, lighting and especially sound. By the time Lord Home came for Speech Day in 1982 and unveiled two plaques, the one commemorating the visit of Princess Margaret on the occasion of the School's 400th Anniversary now re-cut and in a new setting, the other recording the Restoration of the building, the Hall had become part of a pattern of life, the focal point of a School with an ever-increasing range of buildings and a kaleidoscope of varying activities.

The restored Main School Building won two national commendations: one from the Civic Trust and another from *Education* given in conjunction with the Royal Institute of British Architects. Sensitivity of design, craftsmanship and the additional use of space were especially noted, with pride of place being allotted to the Hall, 'a symbol of resurrection'. The building had been viewed soon after occupation when the circulation areas appeared somewhat 'hard, bland and featureless', but pictures, many of them donated, soon appeared in the Lower Foyer and in corridors, and these were to mellow the overall impression.

The Recreation Centre had, meanwhile, also been opened and the Theatre completed. Planned before the fire, it had involved yet another sub-committee of Governors, Officers and Staff and a further run of building, planning and financial problems. But the finished structure was to make a significant contribution to the sporting and cultural activities of the School. It became the focal point for the physical education of all boys from the age of seven years upwards. The Sports Hall offered facilities for practising and sharpening the skills of all the major games while a list of other sports included squash, badminton, gymnastics, karate, trampoline, weight training, basketball, volleyball, table tennis, swimming, water polo and fencing. There was also an area for art exhibitions and a kitchen for cookery classes. The facilities were available to boarders in the evenings after preparation and at weekends.

With the Theatre, drama also became an ever-increasing part of the educational scene. Acting and play directing involved boys in all their associated activities such as lighting, making of scenery, even costumes and stage management. Though designed primarily for school use, the Centre has by degrees taken on a wider role. It is currently used by a wide range of local sports clubs and associations, by staff, Old Bedfordians, parents and friends of the School. The Great Hall too has been used for County and District occasions, for Bedford Music Club concerts and for a variety of public as well as private functions. Both are forging a fruitful relationship between the School and the community at large.

With the Recreation Centre and Main School Building combined, it could be seen that nearly half of the School's facilities had been newly built or reconstructed. Other parts of the Estate had now to be progressively upgraded. An RE department had been omitted from the Main School. The Staff Planning Committee originally suggested a site in 37 De Parys Avenue but cost and convenience led to a more central position in 6 and 7 Burnaby Road next to the Phillpotts Gates. Here plans drawn up by the Trust's Surveyor redesigned the interior to provide two classrooms, a seminar room and Chaplains' studies within one self-contained area. An enlarged School Shop and new premises for the Mothers' Guild and the School Archives were skilfully fitted into other parts of the building.

At the same time the offices of the Old Bedfordians Club were re-sited above the Bursar's offices in Glebe Road. The latter were not moved but retained their comfortable informality, commanding the entry of goods and services to the Dining Halls and the School Estate. Some would perhaps have preferred one new and centralised administrative block for the management of the School Estate, but distributed at the key entrances and at easy distance from the Main School Building they were to service in close association the many functions that any contemporary organisation is expected to provide without that feeling of compression that can so frequently become one of the features of modern office planning.

1983 also saw the completion of plans to refurbish the Memorial Building. An expanded Library had been excluded from the Main School, owing to lack of space. All had recognised, however, that a Library and Reading Room were significant focal points in the academic life of a school and Arup Associates were given the task of completing their work by achieving an

Recreation centre activities: The climbing wall *(S. B. Stearns O.B.)*, swimming pool and a gymnastic display in the Sports Hall. *(Morley Smith)*

improvement in this area. Once again virtually all hopes were realised in the imaginative and sensitive designs of the Architects. In the Lower Library the book stack was doubled to provide space for up to 15,000 volumes in skilfully constructed T-extensions built in oak to match the original shelving. Small tables within the newly formed alcoves allowed a few to work in effective isolation. The older presentation furniture, tables and chairs, so keenly promoted by Eugene Rolfe fifty years earlier, were moved upstairs into a redecorated, carpeted and brightly lit Memorial Hall. This became a Sixth Form Reading Room as well as the scene for termly Governors' meetings. The upper landing was turned into a reading area with easy chairs, tables and magazine racks for newspapers and journals. The small study rooms were left for the examination of filmstrips, micro films and archive materials. A librarian's office and reception desk on the ground floor completed the new design. It was opened officially by Mr H. P. Shallard in his last year as Chairman of the School Committee. His years as Chairman, years of quite exceptional service, had witnessed the second great transformation of the School Estate this century.

Fifty years after its construction it was clear that the Science Building was also in need of renovation. Milne's original had served the School well. It had been extended in the 1950s thanks to a grant from the Industrial Fund, and the addition of the Lucas Wing for Biology had been the last major building project to be completed before the fire. It was now the turn of the Chemistry and Physics Laboratories and Lecture Rooms to be updated to cater for new subjects and a more contemporary teaching approach. The familiar pattern of sub-committee and feasibility study with briefs prepared by Heads of Department was accompanied by the usual cost controls. The Laboratories were restructured over one year in a four-phase scheme allowing the School use of sufficient space for teaching purposes during the contract period. The old Lecture Theatre, whose wooden benches had begun to splinter, was divided horizontally with a new split-level Laboratory and Lecture Area below and an additional Library and Study Area in the roof space. Other rooms were remodelled to allow for an additional Electronics Laboratory, improved study areas and more flexible use of teaching facilities. Modern safety regulations required separate outlets for every fume cupboard and rows of futuristic chimneys ranked like so many space invaders were lined up on the roof, fortunately invisible from the School field! Otherwise it was once again a matter of redesigning an interior while keeping the general appearance of the northern approaches of the School Estate intact, an exercise well realised by the Architects, Feilden and Mawson, who had also been responsible for the construction of the Lucas Wing.

New extensions were also planned for the Workshops. These would extend into the parking spaces to the north of their existing premises and also allow for the consolidation of the Geography Department in the Wells Building. Future plans include the upgrading of the Howard Building and

The Great Hall south colonnade with the Commemoration stone in position. *(Arup Associates)*

for feasibility studies for the Inky and the southern end of the School Estate. With landscaping and replanting imaginatively planned by the Bursar, it will mean the gradual enhancement of the entire area, the removal of several eyesores and the elevation of the 'southern quadrangle' to match the extensive improvements on the northern side. Already the Recreation Centre has become the backdrop to a row of sparkling silver birches, and as the shrubs grow along the south front of the Main School Building the ideals of Planners and Architects can be seen coming to fruition. The ambitions of Grose-Hodge and his successors, the sensitivity of O. P. Milne, the projections of Feilden and Mawson, the artistry of Arup Associates fused into a balanced and coherent design.

But, critically, what of the pupils? How do they fit into this new world? Is it too exposed, too public for their secret ways, too well programmed and patterned for individual and indigenous growth? Care has been taken to sustain and increase the School's small islands of enterprise. The Printing Club has spread to fill the old ground floor of the 'monkey cage', art studios have increased in Howard's, the wine and beer makers have secured new premises in Brown's, one of the new day boy houses sited at the entrance to Burnaby Road. The Computer Room has acquired more and more hardware to occupy the initiated through all hours of the day, much of the night and also during the holidays. New project rooms are planned in the Workshops and Laboratories to encourage individual enterprise during the week and at week-ends as far as safety factors will allow. Possibly the mountaineers feel confined to their climbing wall rather than ranging freely and rather expensively as was their habit over the roofs and buildings. It is perhaps increasingly hard to find odd corners which can be used and adapted to meet the hopes of boys or members of staff wishing to expand existing activities or to inaugurate new ones. But in a world of multiplying crafts and skills overall planning and organisation have become essential to ensure that everything is fitted in.

Nevertheless, there is no doubt that the new facilities have provided fresh opportunities for boys to acquaint themselves with a wider spectrum of activities than ever before. Touring companies provide regular shows in the Theatre; solo artists and orchestras perform in the Great Hall. The numbers prepared to appreciate them are still rather small. But education proceeds by involvement with a rising tide of dramatic productions. Plays have been put on by the Staff, by the Modern Languages and Classics Departments, by forms and individual groups as well as the full-scale presentations at Senior and Junior level. Bands, orchestras and individuals from the Inky to the Upper School perform in the Great Hall, gaining each year in confidence and technical control. The Festival of Music proceeds annually with additional performers and performances. As facilities for Art and Workshops expand, so opportunities for a wider distribution of skills and increased individual contributions should emerge.

Sport has, at first sight, also seen the advance of the big battalions. In 1983

The new lower foyer with commemorative plaque. *(Morley Smith)*

the Rugby Club played 117 inter-school games at 21 different levels involving 340 boys. Thanks to the Sports Hall, cricket and hockey can be played all the year round. Training for oarsmen has spread with the aid of the Weight Training Room into the Christmas term. The all-weather permaprene surface allows games to continue in virtually all conditions. While fencing and fives survive, tennis, squash, swimming and athletics flourish and have taken over the old, more specialised, skills of gymnastics and boxing displayed in the once popular Assaults-at-Arms. More boys are catered for. But what of their quality? Allowing for the few outstanding sportsmen, the average may be classified as slowly rising, but only when the commitment is consistent and the coaching formative and carefully programmed. Standards rise, but the starting points remain fixed, the raw youth eager but quite unprepared. Will new facilities help to bridge the gap or merely over-extend human energy and resources? Is too much expected too soon?

Training for new technologies and management skills has been one of the innovations of the last decade. All are introduced to the computer, even members of Staff! Follow-up courses can be taken at a general or more advanced level. New electronics programmes are being incorporated at various points in the curriculum. Boys may be seen in the afternoons laying cables to connect the audio-visual facilities to classrooms and work areas, clearing the ducting with draw wires and springs, studying new computer and word-processing facilities. A number sit with masters on a wide spectrum of committees discovering the inevitable frustrations of management and the obstacles to change. Does this breed cynicism and sap the vitality of youth? Perhaps, but it is difficult to see how else the young can be made aware of the difficulties ahead and acquire a realistic understanding of the ways in which contemporary society tries to solve them.

131

There are no signs of loss of vitality or initiative in the Combined Cadet Force or Community Service sectors. Adventurous training, self-reliance exercises in Yardley Chase, camps and courses all flourish, allowing many a cadet to gain experience and confidence in unexpected quarters. New skills are learnt and applied from the possibilities of survival to greater control of sail, rifle or sophisticated communications systems. If the Community Services feel lacking in skill, the gaps may be made up by fuller use of the instruction available in Projects, Courses and elsewhere. If boys become too individually minded they can be harnessed for wider use within the community.

And what of academic work? What of that slow and often painful process by which the mind achieves an element of understanding and gains confidence it its own judgements? It can neither be disguised within a programme of colourful activity designed to satisfy the prospect of some future 'leisure age' nor be lodged in a convenient computer for subsequent and more general use. It has somehow to be engendered in every pupil; but how? Here the balance between instruction and individual commitment is possibly the most difficult to achieve. Much of the process is both invisible and unquantifiable. That desire for instant knowledge and immediate certainty can all too easily undermine the acquisition of genuine confidence and personal understanding. The pressure of examinations and fear of the half-known make many clutch at the present rather than explore what appears to be an all too uncertain future. To achieve a balance between being taught and discovering how to learn is not easy. The layout of the new departments with their teaching rooms, new laboratories, Staff bases, libraries and study areas was designed to provide a physical framework to assist these positive processes. There are signs that some are being used with growing discernment. But new habits are not formed or acquired overnight. Pupils will always seize on teachers' notes rather than make their own. The future may well test out the leaders as well as the led.

Much depends on the human equation: the relationship between staff and boys, old and young, master and pupil. The fire and rebuilding drew all together in a combined effort to ensure survival and achieve reconstruction. The Chapel provided an appropriate refuge in the hour of need – a not infrequent occurrence in English history! The 75th Anniversary of its dedication on 11 July 1908 was celebrated in the last week of the Summer term 1983 with a Flower Festival organised by the Mothers' Guild and a memorable lecture on the Architect, G. F. Bodley, given by S. E. Dykes-Bower. It was visited by over a thousand Bedfordians, many of whom expressed considerable surprise at its existence! And what would it mean for the future? The church is familiar with its contemporary predicaments. The communication of values and infinite perspectives is not easy, yet its significance for the young should not be underestimated. Those in revolt against society and sometimes against themselves must have something to lean on. The staff must comfort still.

Bridging the generation gap is not a problem confined to schools but absorbs the home, the community and the whole country. It is possible that schools may appreciate more of its problems, but only because they become involved in so many of its dimensions. The dayboy's difficulties in adjusting between school and home life, disciplines and freedoms, the boarder with his abrupt transition from the holiday to the term and back; both require the advice and guidance of sympathetic counsellors familiar with all estates and conditions of men. Here the new dayhouse units are slowly playing a constructive role. In boarding houses new senior study facilities and middle common rooms allow each generation to evolve. Informal assistance from housemasters, tutors, teachers and careers advisers helps to assess the obstacles and point the way. Not that all will be satisfied. But most should have a life-line through the labyrinth.

Both Sir Philip Dowson and the Chairman of Governors had referred in the dark days of March 1979 to a pilgrimage of reconstruction. They were in the town of the greatest English pilgrim and Sir Philip would in effect have to lead them through the Slough of Despond and help to overcome Castle Doubting and defeat the Giant Despair. The heritage and the future of the School were in his hands. But what of those who followed after? A flame has to be kindled in the mind of each pupil if the future is to be given meaning and if the present is to realise any useful purpose. Has too much been provided? Is too much taken for granted? The best creative work has not infrequently come out of huts and attics, and the School had its share of these in the pioneering days at the turn of the century! Is there now too strong a temptation to linger in the House Beautiful or in Vanity Fair and to avoid the more challenging tests of the Wilderness and the Valley of Humiliation? Life needs its oases among the shifting sands; islands of relative calm in the cross-currents of a contemporary world. Where else is the sown seed to germinate, to grow and to mature?

Schools can only be points of departure bequeathed by those who have gone before for the benefit of those who must follow behind. In their variety, their balance between discipline and order, instruction and opportunities for experience, they have to reflect the many facets of contemporary society and prepare the young for their roles and relationships within it. They have also to sustain the underlying threads of thought and action which have prompted the emergence of civilised man during the course of the last three thousand years through accuracy of mind, resilience of body, artistic response and spiritual awareness. The balance between idealism and realism is hard to achieve. But man must continue to aspire if he is to survive and evolve. Bedford has for over a century been an aspirant school, hopeful of ideals, tempered by realism and essentially practical and purposeful in design. Those who planned and rebuilt it after the fire had a unique opportunity with others of their generation to contribute towards its evolution. They contained possibly as wide a variety of talent as one is likely to find in one place at any one time. The future will surely vindicate their work.

Epilogue

I congratulate the three compilers on a fascinating and most readable account of the Great Fire at Bedford School. It is an excellent study of a historic period in the life of the School. We should also remember that as members of the Staff Planning Committee, all three played a notable part in the planning of the 'new' School. This book is therefore an important source for future historians, but it is not a definitive history. We were all too closely involved between March 1979 and September 1981 to be able to view the events with sufficient objectivity and lack of emotion, or to expect to get agreement on the emphasis and balance of what actually happened.

When I was asked to write the Epilogue I felt this dilemma most acutely. I read and reread the proofs of the book and consulted the bulging files which my P.A. and I kept throughout the period. Most of the story was there, but somehow it was not totally satisfactory. When I mentioned this to the Head Master I was relieved to find he felt the same. We had both enjoyed reading the proofs but did they tell the 'whole story'? And then we discovered that, between ourselves, the Headmaster and I shared many points of view but also differed slightly on others. Two Governors, Gilbert Beazley and Patrick Shallard helped to resolve our anxieties by remarking that it was still too early to evaluate, we were all too personally involved.

So this Epilogue must be subject to the same limitations as the rest of the book. It should be read as the view of one person who was continuously at the centre of things.

The Great Fire at Bedford School has influenced the lives of several thousand people all of whom played some role at the time and during the reconstruction. Staff, boys, parents, local authority officers from several departments, the police, the fire brigade, the loss adjustor, the Phoenix Insurance Co., the team set up by the architects, the Contract Managers and their sub-contractors, the Commissioners of Customs & Excise, lawyers, surveyors, Members of Parliament, the press and television, the National Westminster Bank, Harpur Trust staff, old Bedfordians, friends of the school, the list grows and grows, and all had 'their exits and their entrances'.

I have not yet mentioned the most important group of all. It is said that President Truman had a plaque on his desk in the Oval Office – 'the buck stops here'. The ultimate responsibility lay with my colleagues, the Governors of Harpur Trust, supported magnificently by the Headmaster. Upon their shoulders alone rested the burden of 400 years of history, the immediate catastrophe of the Fire and the future destiny of the School. This book tells a success story and the plans which provided the foundation and framework for that success were completed and communicated to those who executed them within 14 days of the Fire.

Three colleagues set out those plans in the first 72 hours. The Headmaster, Patrick Shallard as Chairman of the School Committee, and myself as

Chairman of Governors. On Sunday, 4 March, the Headmaster had already dealt with the immediate problems and he and Patrick Shallard began to outline priorities. The next day the Headmaster and I met and agreed the strategy. The note reads:

Priorities agreed on 5 March by Chairman of Governors and Headmaster
1. School will function normally until end of term.
2. Main building must be made secure.
3. Temporary accommodation must be planned, sited, ordered and erected before beginning of next term, 18 April.
4. Extent of losses to school and individuals must be quantified as soon as possible.
5. Negotiations with insurers must begin.
6. Main school building must be restored or rebuilt by September 1981.
7. All offers of help and expressions of sympathy must be dealt with promptly.

On 17 March at a meeting of the whole governing body, we were able to report that under each heading details had been worked out, jobs allocated and satisfactory progress had been made. At this stage two items were of particular importance, the temporary classroom accommodation, and the restoration or rebuilding by September 1981.

As to the temporary accommodation, my predecessor, Sir John Howard (OB), together with the Headmaster, Bursar and Surveyor, sent out specifications and obtained quotations for 22 temporary classrooms within 7 days. By 14 March, terms had been agreed with the Contractors. Erection began on 20 March and handover took place before the next term began on 17 April. Sir John also kept an eye on the possible resale of the classrooms which the Bursar at the end of 1983 achieved with great success.

To restore or rebuild? A firm decision was required as quickly as possible. For a few days one or two of my colleagues were undecided; the remainder wished to retain the shell provided it was structurally sound. From the out-set I wanted the familiar north facade to remain as a reassurance to old Bedfordians and future generations. All of us wanted to improve the South Elevation and redesign the interior to bring the facilities into the 21st century. On Tuesday, 6 March, the Headmaster and I gave our views to the whole of the staff and told them that they would have the responsibility for preparing the educational brief which the architects must interpret, and that the main school building would be opened again in September 1981. Although there may have been some at the meeting who had doubts I know that when they returned for the next term 6 weeks later those doubts were dispelled.

On 17 March, all the Governors of the Harpur Trust met and the Arup team, led by Philip Dowson, was appointed. This led to the remarkable partnership of Governors, Head Master, staff planning committee and architect which achieved all the objectives: restoration of the facade, magnificent improvement of the South Elevation and brilliant redesign of the interior. All the fundamental decisions were boldly taken by the Governors within 6 weeks of the Fire. On the financial question, Patrick Shallard and I agreed with the school committee that this must be left until two matters were settled – the educational brief and the insurance claim. Patrick Shallard wanted these to be presented simultaneously together with a preliminary costing of the brief. On 17 July, the architects presented their plans together with preliminary costings to the School Governors and the insurance claim had been settled. When the full governing body met on 26 September to discuss (and approve unanimously) the final design, a cost limit of £3.5 million had been set and this was achieved when the final bills came in nearly three years later. Of course, there were moments of doubt, for example, when the question of VAT assessment of £430,000 emerged, owing to an unexpected review of the VAT rate, but my colleague, Gilbert Beazley together with the Finance Officer undertook the negotiations with the Customs & Excise and, although the subsequent story of the VAT assessment merits a little book of its own, we won our claim. The building was not liable to VAT. As to the remaining questions of finance, the Chairman of the Harpur Trust Finance Committee, Ronald Gale, handled the question of the structuring of the loan required and the formal arrangements with our bankers. Reports to committees and discussions within committees on financial matters took place from time to time and different views were aired as to the feasibility of raising by appeal or selling property but I do not recall any difference of view between Patrick Shallard, Ronald Gale and myself and once the financial target had been set the strategy did not change.

These are examples taken from my files which illustrate the importance of declaring our objectives at the outset, then defining everyone's responsibilities and sticking to the policy. I cannot pay sufficient tribute to the way in which the School Governors led by Patrick Shallard monitored the scheme, reporting regularly to the rest of the governing body. It should be remembered at the same time Bedford School was in the middle of building a large sports complex and theatre within 75 yards of the main building and our longest serving Governor, Arthur Jones, monitored the progress of this large development. Meanwhile, the rest of the work of the Trust went ahead with development plans at our other three schools, Alms Houses, and on our London Estate. The governors looked squarely at all these plans and confidently decided to go ahead with them.

The foundation stone ceremony, the topping out, the service of dedication were all happy and emotional occasions. During the whole period messages of sympathy and offers of support came from all over the world. The expressions of support and practical help from so many comforted and inspired us. Now every time I walk into the stupendous Great Hall and look out over the school field through the mullion windows I give thanks that as a Governor and old Bedfordian I was so fortunate as to be allowed to play a part in handing on to our successors such a marvellous building.

Anthony Abrahams

Appendices

**Calendar of Events during the
last fifteen days of Term**

Saturday March 3: Fire breaks out at an estimated 11.45 pm in the Main Building; the cause remains to be established beyond doubt.

Sunday March 4: 12.01 am County Fire Headquarters receives the first alarm call (a total of 22 further calls were received); 12.06 am the first fire tender arrives; Chief Fire Officer Reginald Haley takes personal charge of the operation; 13 pump tenders, 3 platform engines, 1 turntable ladder are used to fight the blaze; 3.20 am Chief Fire Officer declares the blaze under control.

Assessment later in the day reveals the extent of damage: 30 classrooms, 3 Staff rooms, the administrative offices, the Head Master's study, the language laboratory, the library annexes of the Modern Languages, Classics, and History Departments, in the Great Hall all panelling, chairs, portraits, the lectern, the chairs 'of state', the organ, the Bechstein grand piano, the hammer beams, the pitch pine ceiling and the entire roof and spire destroyed. Contingency plans for the resiting of classrooms throughout the School estate are prepared. Notices indicating location of Monday morning assemblies are posted at the School Gates. Work begins on setting up the administrative offices in 6 Burnaby Road. A letter from the Head Master to all parents is prepared. A filmed report on the Fire is shown on ITN News at 6.45 and 9.45 pm.

Monday March 5: The Head Master addresses the VIth, Vth and Remove forms in Chapel; the Vice-Master takes IVth form Assembly in S18; Mr Allerton takes Lower School Assembly in the Gymnasium. Normal timetable begins with Period 1 in the new locations. Classrooms are found in Pemberley, Farrar's, Talbot's, Burnaby, the

Pavilion, the Rifle Range. A Statement to all Governors and other interested parties and the Letter to Parents are despatched. A full Staff Meeting is held. Peter Dann and Partners (consulting engineers) are asked to prepare a preliminary report on the state of the structure for safety reasons. Press reports appear in the Daily Telegraph, Daily Mail, Sun, and Guardian newspapers; coverage also on TV in Anglia local news and BBC Look East. The Fire Officers remain in attendance throughout the day.

Tuesday March 6: 8.32 a.m. the Fire Officers declare 'incident closed' and stand down after 56 hours' continuous attendance. Of five inter-school Hockey matches played in the afternoon four are won, with one drawn. An ITV film crew spends two hours filming for the evening news on Anglia. One member of the film crew subsequently entered his son for the School. The first of the three Parents Evenings of the week is held as scheduled, but in the Dining Halls.

Wednesday March 7: Preliminary report received from consulting engineers. Surveyor orders erection of chainlink fence with padlocked gates so that the Main Building can be isolated. John Wilmott Ltd is asked to secure the east and west gables. Chairman of School Committee is asked to see to the provision of temporary accommodation to be in operational use for April 18.

Friday March 9: Bedfordshire Times publishes a full report of the Fire together with many photographs. Quotations are sought for temporary classrooms.

Monday March 12: a full week of House Plays begins with Redburn's production of *The Importance of being Earnest.*

Tuesday March 13: The Bursar, the Surveyor and Mr Wood visit Austin Hall Ltd.

Wednesday March 14: Order sent to Austin Hall Ltd for the supply of 11 units containing 22 classrooms with cloakroom and storage accommodation. School Committee accepts the recommendation of Head Master and Staff that the units should be sited to the west of the Chapel and to the north of the Lucas Wing, thus making the Chapel a new central focus for the Upper School. A Staff Planning Committee is nominated under the chairmanship of the Vice-Master to prepare an outline draft of accommodation requirements and priorities for the reconstruction of the Main Building.

Saturday March 17: The Governors select ARUP Associates, Senior partner Mr Philip Dowson CBE, to prepare plans for the reconstruction of the Main Building. ARUP Associates have been involved in the Maltings at Snape, the Theatre Royal Glasgow (for Scottish Opera) the Royal Opera House, Ampleforth College, and in buildings in the Universities of Oxford, Loughborough, Sheffield, Cambridge, Surrey, and East Anglia.

Monday March 19: The last day of Term: in Final Assembly the Head of School presents a giant good wishes card, signed by all boys in the Upper School, to the Head Master as a token of admiration for his leadership during the last fortnight of Term. Cross Country Races are held as Austin Hall come on site to prepare drainage and electricity supplies for the temporary classrooms.

During the Holidays: work on fencing off the Main Building, boarding up the ground floor doors and windows, securing the gables with scaffolding and timbering was completed. An offer from the Local Education Authority for the loan of classroom furniture was gratefully accepted and this was delivered on April 11. The shortfall of furniture was ordered from PEL Ltd on March 26 and delivered on April 13. The classrooms were completed by Austin Hall Ltd and taken over by the School on April 17. Work on the three major developments on the School Estate continued unimpeded. These are the provision of maintenance staff quarters adjacent to the Fives Courts, the alterations to Talbot's, and the Recreation Centre. This latter project, designed by Mr R. S. Hollins, will provide a Sports Hall, a six-lane 25m indoor pool, four squash courts; the Gymnasium will also be converted into a Theatre which can be used for Lower School Morning Assembly.

Ordeal by Fire 1945

BY OUR SPECIAL CORRESPONDENT

From *The Ousel* April 4, 1945 page 2.

On the last day of February, at about a quarter to seven, a small boy was making his way back from a School engagement to his home near the School. When he arrived, he remarked to his father: "There's a lot of smoke coming out of the School, and it's not coming out of the chimney." His was the first warning of the fire which so nearly burned the main building of Bedford School to the ground that Wednesday night.

His father rang up the Fire Station; and by the time that the present writer reached the School some ten minutes later, the Fire Brigade was already in action, the hoses had been brought up to the top landing, and the firemen were trying to locate the fire with the assistance of three Monitors and Sergeant Jackson, who was performing feats of more than gymnastic ability in the roof. Undeterred by a Monitor who tried to extinguish him with a "Mini-max." your correspondent took a look round. Some rather acrid smoke was billowing about the top landing, and making his way through it, he entered each of the classrooms one after the other. Here there was nothing to be seen—no flames, no smoke even; everything neat and orderly as it had been left a few hours before. Then he noticed a light, somehow different from the lingering yellow daylight; and looking up, his eye was drawn to the ventilator in the classroom ceiling. Through it could be seen the flames of a roaring furnace. It was the same in every classroom along the top gallery. Starting between the roof and the ceiling of C5, and fanned by a moderate west wind, the flames had quickly made their way underneath the roof as far as C9, without anything but a little smoke to show what was happening. If the fire had occurred at night, it could hardly have been discovered in time.

Even so, the situation looked well nigh desperate. By the time that the hoses discharged their first jets upon the blaze, the beams supporting the roof of the south side were ablaze from end to end. It seemed only a matter of minutes before the flames would spread to the roof of the Great Hall; and if that happened, nothing could save the School. The dry pitch-pine of the roof would go up like a torch, till the "copper spike" crashed down, to hurl burning fragments among the chairs and to fire the panelling of the walls.

What saved us? Two things only: the solid construction put into their work by the builders of 1891, and the skill and bravery of the fire brigades. These soon began to arrive from neighbouring towns till there were nine in all; and by the time that the horrified spectators saw great bursts of flame leaping up from the roof (through holes, incidentally, made by the firemen for them to leap through) an overwhelming mass of water was being pumped on to the fire by the hoses; and it was gradually realised that the blaze of light with which the School was surrounded came less from the conflagration than from the searchlights which, raised to an incredible height at the end of tapering steel antennae, were directing the firemen at their work. By nine o'clock the fire was out.

II

Ordeal by fire was over, but there remained an ordeal only a little less grim, by water. There was water, water everywhere. It poured through the floors of the damaged classrooms, through the ceilings of the rooms below, and again, in sheets and cascades, through them into the ground-floor classrooms. It came down the west staircase in a foaming torrent and swilled about ankle-deep in the vestibule, corridor and porch. The Head Master's Study, the Office, and the Bell Room were three lakes. Though saved from burning, the school was drowned.

But we had not just waited to see what would happen, and it is not too much to say that before the water could get at it, everything of value that could be moved, had been moved to safety: the records and papers from the Office and Bell Room, the books and furniture from the Head Master's Study, the instruments and apparatus of the B.B.C., including two grand pianos, from the Hall, the pictures from the galleries, the form libraries from the classrooms. All this was saved by bands of Masters and boys; and your correspondent noticed parents, Old Bedfordians, and friends of the School among the helpers.

But much could not be moved, the panelling, for instance, with which years of work and a thousand generous gifts have equipped our Hall. That all this is hardly damaged beyond what a little polish will restore, we owe above all to the work of the Fire Brigade, who toiled all night damming the flood with sandbags and sweeping it out of the School. For this, and for all their skill and gallantry, we offer them our heartfelt thanks. They are kind enough to say that our A.R.P. arrangements helped them; extra ladders were in place, leading where it would otherwise have been hard to reach. The positions of water, gas and electricity were clearly marked. And, greatest good fortune, the Brigades of the district had already practised dealing with an outbreak of fire on the School roof, which they took as their "scheme" for a special field day held on the spot last summer. If the fire had occurred in peacetime, the damage would have been easier to repair; but, in fact, it would have been irreparable.

III

The boys of the School, having had no orders to the contrary, assembled as usual for nine o'clock Prayers next morning. Some had not heard that there had been a fire. Others had heard that the School had been burned to the ground. Others, again, who had only left the Great Hall a few hours before, with water cascading down from the galleries, the chairs in a heap in the middle, and the platform furniture removed for safety, may have dreaded to think what it would like in the cold light of "the morning after the night before". They found it exactly as usual. Not a chair was out of place. The fact is that a Monitor and a squad of twenty boys had arrived early and had put the whole Hall straight in half an hour.

After Prayers (at which the singing of "Oft in danger, oft in woe", to our own lively tune, was noticeably hearty!) we received our emergency orders. In brief, all School engagements except school periods were to be carried on as usual; that meant games, music lessons, School societies, the Gymn. Club, Scouts—everything, in fact, except work. And work was to start again with a full time-table on Monday morning. In the mornings of the intervening days, the Fifth and Sixth Forms took it in turns to work at salvage, and they worked with a will. Within twenty-four hours of the fire, the top floor was cleared, tons of sodden debris were carried down in buckets and dumped outside, instead of being left to hold the wet and pass it on to the ceilings below. Damaged desks and furniture were salved or, with greater gusto, thrown out of top floor windows. Sawdust was laid down where water had been inches deep the night before. Every door and window was opened; and through them, as well as through holes in the roof, the brisk sun and wind of perfect March weather poured in to do their healing work. When the School reassembled on Monday morning, though five class-rooms along the top gallery were out of commission, every form had a room of its own and not a period of work has been missed since.

And now the School is looking very much itself again. The damaged roof is covered with tarpaulins, the scorched timbers have been hacked away and the loose tiles removed. We are wind-and-weather-proof—unless the equinoctial gales decide otherwise. We hope to get some of our damaged classrooms back into use next term, and the rest by the end of the summer holidays. "If hopes are dupes, fears may be liars"—so why not go on hoping?

IV

Meanwhile many friends had been sending in offers of help—from the Mothers' Guild and the High School to single individuals, like the

O.B. who rang up to say that he had heard that the School had been burnt to the ground, and should he come down at once? One and all we valued them, for they spoke of warm hearts and willing hands. But the only one which, so far, we have been able to accept was that of the High School, which sent round a party of girls on the Friday morning to help in rehabilitating the Bell Room, the Office, and the Head Master's Study, which had suffered most from water, and of which the inhabitants had somehow to live and work in them. The smiling faces of the party were as much appreciated as their hard work, and we shall remember long and gratefully this truly sisterly gesture. A visit at the same time from General Fuller, Deputy Regional Commissioner and the Officer responsible for the fire services of the Area, gave us further encouragement, and the comfortable knowledge that we have a "friend at court".

Letters, too, reminded us of the concern which any misfortune to the School must cause to a large circle of friends. Kind expressions of sympathy arrived from the President, Secretary, and Treasurer of the Old Bedfordians Club, from Field-Marshal Sir Cyril Deverell (ex-President), from the Committee of Victoria College, Jersey, the Head Masters of Bedford Modern School, Rye Grammar School, and several of the schools which we play, as well as from many parents and O.B.s. "I would have walked from London with a bucket of water", wrote one of them, "if it would have done any good," while another expressed what may have been the feelings of many when he wrote: "When I was at School, the roof was just a roof, and there was nothing more to it; but now I find that it has taken on a new significance altogether."

So the fire has brought us, in all these expressions of loyalty and concern, good as well as ill. What are its lessons for us? A better understanding, perhaps, of how much we depend upon public services like the Fire Brigade, which we ordinarily just take for granted; appreciation of the importance, as well as the nuisance-value, of A.R.P., for which we had cause to be thankful on the night of the fire, and of our routine fire-drill, which would, we believe, have taken every boy to safety had the School been full, instead of empty, at the time; and a warning of what fire can do, in case we should be tempted to forget these things.

And the cause of the fire? Official investigation has so far yielded no answer, but rumour is both busy and confident. It was anything from a German aeroplane crashing on the roof to a "judgment" on the School authorities for not having given the School a half-term holiday!

Firelight on the Main Building

The fire at Bedford School has revealed certain virtues of the Victorian era. There is a tendency of the present generation to belittle the ideas and deeds of their great-grandfathers, but an inspection of the damage done to the roof and upper works of the School shows that the craftsmen of the 1890's could at any rate build soundly and solidly.

We may criticise the Victorian conception that a scholastic building must be clothed in the gothic style of architecture. Brick gothic with sash windows behind stone mullions looks to us a little absurd, and hard-pressed machine-made bricks and tiles are unsympathetic materials at any time, and quite incompatible with a style of building which was the outcome of an age in which the machine was unknown. There is, however, much to be said for the practical layout of the School, with its Great Hall and its spacious, airy and light classrooms ranged around it, with an economy of corridor.

The builders of the main school certainly had none of the modern notions of speeding up building on prefabricated lines, or of reducing their materials and man-hours to a minimum.

One is struck by the real solidity of the work; for example, the brick-work of the top floor is 18 inches thick. The plastering of the ceilings is carried on expanded metal. This must be a very early use of this substitute of metal for the age-old method of using rent wooden laths to hold the plaster.

Between the roofs of the classrooms on the southern side of the building and the Great Hall there is a broad gutter, well constructed with lead. The inner roof of the classrooms, sloping to this gutter, is flatter in pitch than the roofs which are seen from the ground. Slates have been used to cover this flat pitch. Although this slating cannot be seen (except from an aeroplane), the best thick slates from Westmorland have been used.

The School moved into its new building in 1892. The cost of the building (£25,000) compared with the figures of building costs which rule to-day, would seem incredibly small. However, this relatively modest cost was certainly not achieved by any resort to flimsy or shoddy construction.

O.P.M.

The Governors of the Bedford Charity (Harpur Trust)

Life Governor Emeritus:
Sir John Howard D.L., D.Sc., F.I.C.E.

Ex-Officio (1)
The Member of Parliament for the Bedford Constituency:
T. H. H. Skeet M.P., LL.B.

Nominated (12) (for 5 years)
By New College, Oxford:
Dr D. F. Mayers M.A., PH.D.
W. M. Pybus M.A.
By Cambridge University:
Dr. M. M. Bull M.A., M.D.
By London University:
Prof. C. J. Constable M.A., B.Sc., D.B.A.
By the Permanent Staffs of the School:
Bedford School: A. C. W. Abrahams M.A.
Bedford Modern School: J. I. H. Hadfield B.M., F.R.C.S.
Bedford High School: Miss E. Alexander O.B.E.
Dame Alice Harpur School: Mrs O. J. J. Rogers M.A.
By Parents of Day Scholars of the Schools:
Bedford School: S. J. Rayner
Bedford Modern School: Professor D. W. Saunders Ph.D., A.R.C.S., F.I.P., F.P.R.I., F.I.M.
Bedford High School: Lady C. Chilver
Dame Alice Harpur School: Mrs R. H. Welch

Co-optative (4) (for 5 years)
G. C. W. Beazley F.R.I.C.S.
R. G. Gale
Arthur Jones F.S.V.A.
H. P. Shallard O.B.E., M.A.

Representative (13) (for 4 years)
North Bedfordshire Borough Council
D. Bygrave
Mrs A. N. Polhill
Lt.-Cmdr. R. A. Whittingham D.L.
County Council of Bedfordshire
F. D. Bailey B.A
L. G. Bowles D.L
I. L. Dixon
F. L. Edwards
B. K. W. Gibbons B.Com., A.C.I.S., A.C.C.A., A.M.B.I.M., A.C.A.(NZ)
P. J. Gordon
A. P. Hendry
F. S. Lester
Mrs G. M. Rose
Miss M. C. Shepherd M.B.E.

The Restoration Committee

Mr A. C. W. Abrahams (OB) (Chairman), Chairman of the Harpur Trust
Sir John Howard (OB)   Life Governor Emeritus*
Mr R. G. Gale   Chairman of Finance Committee
Mr H. P. Shallard   Chairman of Bedford School Committee
Mr A. A. Jones   Chairman of Estate Committee
Mr G. C. W. Beazley   Vice Chairman of Bedford School Committee

The Bedford School Committee

Mr H. P. Shallard*   Chairman
Mr G. C. W. Beazley   Vice-Chairman
Dr M. M. Bull*
Lady C. Chilver*
Professor J. C. Constable
Mr P. J. Gordon
Mr A. P. Hendry
Mr W. M. Pybus
Mr S. J. Rayner
*Members of the Restoration Sub-Committee

The Bedford School Staff Planning Commitee

Mr C. I. M. Jones   Head Master
Major D. R. Mantell (OB)   Bursar
Mr M. E. Barlen   (Vice Master), Chairman
Mr T. C. Allerton   Headmaster of the Lower School
Mr P. R. O. Wood   Usher
Mr M. P. Stambach (OB)   Head of Modern Languages
Mr D. P. C. Stileman   Head of English
Mr M. J. Rawlinson   Head of Mathematics
Mr D. W. Jarrett   Secretary

The Restoration Finance Committee

A. C. W. Abrahams (OB)   Chairman of the Harpur Trust
Col. R. R. St. J. Barkshire   Chairman
R. G. Dewe   Parent
M. J. Ferdinando   Finance Officer of the Harpur Trust
C. I. M. Jones   Head Master
Major D. R. Mantell   Bursar, Bedford School
H. P. Shallard   Chairman of the School Committee
A. G. B. Young   Chairman of the Appeal Committee

The Arup Associates Studio

Sir Philip Dowson   Senior Partner
Derek Sugden   Partner & Acoustics
Charles Wymer   Partner & Structural Engineer
Dick Lee   Structural Engineer
Mike Latham   Quantity Surveyor
Graham Preston   ”   ”
Peter Kelsall   ”   ”   *(part of the time)*

Julian Bicknell	Architect
Michael Lowe	"
Ros Wilkinson	"
David Yearley	" *(in training)*
Keith Barnes	"
Alan Ross	Mechanical Engineer
Stuart Wells	" "
Joe Lewis	Electrical Engineer
Terry Moody	Public Health Engineer
Chris Hay	Interior Designer
John Warman	Administrator

The Laing Project Staff

Tony Legg	Contracts Manager
Bill Critchley	Project Manager
David Pelter	Quantity Surveyor
Kenneth Cole	Resources Manager
Roger Hall	Construction Manager
Graham Jelly	Buyer
Robert Freeman	Construction Supervisor

Clerk of Works

Alan Lamb

Insurance

P. Swann	Manager Bedford Branch, Phoenix Assurance
J. R. Simpson	Loss Adjuster, Thomas Howell, Selfe & Co.

Officers of the Bedford Charity (The Harpur Trust)

R. N. Hutchins	Clerk to September 1979
A. James	Clerk from September 1979
M. J. Ferdinando	Finance Officer
H. T. Inskip	Bedford Surveyor
D. R. McKeown	Development Manager 1979
J. Willis	Solicitor

BEDFORD SCHOOL RESTORATION APPEAL

PATRONS

RT. HON. VISCOUNT BOYD OF MERTON, C.H., D.L.
RT. HON. BARON SOAMES OF FLETCHING, C.H., G.C.M.G., G.C.V.O., C.B.E.
SIR ADRIAN BOULT, C.H.
SIR JOHN HOWARD, D.L.
SIR JOHN BETJEMAN, C.B.E.
SIR PETER PARKER, M.V.O.
SAMUEL WHITBREAD, J.P., D.L.

Appeal Council

Chairman: A. G. B. YOUNG, M.A.

A. C. W. Abrahams, M.A.
   (*Chairman, The Harpur Trust*)
Col. R. R. St. J. Barkshire, T.D., J.P., A.I.B.
T. J. Bedford
Dr. E. V. Bevan, T.D., D.L., M.A., M.D.
The Reverend Canon D. H. Booth, M.B.E., M.A.
The Reverend W. M. Brown, M.A.
J. P. U. Burr, M.B.E.
Sir Henry Chilver, M.A., D.SC.
W. F. M. Clemens, C.B.E., M.C., L. OF M. (USA),
   M.A.
Professor C. J. Constable, D.B.A., M.A., B.SC.
J. G. P. Crowden, J.P., D.L., M.A., F.R.I.C.S.
Professor R. D'Aeth
Richard Dawes, J.P.
Roger Dalzell, M.A.
John Dankworth, C.B.E., F.R.A.M.
N. E. Dudley, M.A., F.R.C.S., F.R.C.S. (ED.)
Dr. G. B. R. Feilden, C.B.E., F.ENG., F.R.S.
Major-General H. R. B. Foote, V.C., C.B., D.S.O.
Sir Bruce Fraser, K.C.B.
Vice-Admiral Sir Charles Hughes Hallett,
   K.C.B., C.B.E., F.B.I.M.
Vice-Admiral Sir Raymond Hawkins, K.C.B.

Brian E. Howard, M.A.
C. I. M. Jones, M.A. (*Head Master of Bedford School*)
General Sir Sidney Kirkman, G.C.B., K.B.E., M.C.
N. Lyster-Binns
C. P. McGinty
Michael Morris, M.A., M.P.
Gerry Neale, M.P.
I. G. Peck, B.A.
Marshal of the Royal Air Force Sir Thomas Pike,
   G.C.B., C.B.E., D.F.C., D.L.
W. M. Pybus
S. J. Rayner
   (*The Parent Governor—The Harpur Trust*)
H. P. Shallard, O.B.E., M.A.
   (*Chairman, Bedford School Committee—The Harpur
   Trust*)
T. H. H. Skeet, LL.B., M.P.
Professor Quentin Skinner
J. G. Vaughan, F.C.A.
and (ex officio)
   The Head of the School
   The Chairman, The Friends of Bedford School
   The President of Bedford School Mothers' Guild
R. C. Williams

EXECUTIVE COMMITTEE

John D. Bardner, B.A.
R. G. Dewe, M.A.
M. J. Ferdinando
A. M. Lloyd-Williams, B.A.
M. B. Maltby, B.A.
Major D. R. Mantell

D. N. Miller, M.A., F.C.A.
A. D. Nightall (*Secretary, The Old
Bedfordians Club*)
D. P. Rogers, O.B.E.
A. L. Wylie, M.C.

Honorary Treasurer: J. P. SOUTHCOTT

Appeal Director (London): J. G. DOUBLEDAY, O.B.E.   *Appeal Director (Bedford):* R. A. FRANKS, M.A.

BEDFORD SCHOOL RESTORATION TRUST

(Registered with the Charity Commission as a Charity)

Trustees: R. R. St. J. Barkshire, T.D., J.P., A.I.B. (Chairman Restoration Finance Committee), C. I. M. Jones, M.A.
(Head Master of Bedford School), H. P. Shallard, O.B.E., M.A. (Chairman, Bedford School Committee—The
Harpur Trust), A. G. B. Young, M.A. (Appeal Chairman)

Bankers: National Westminster Bank Limited, Bedford

THE RESTORATION OF BEDFORD SCHOOL

Consulting Architects: Arup Associates (Sir Philip Dowson, C.B.E., Senior Partner).
Principal Contractor: Laing Management Contracting Limited (T. W. Fleming, Managing Director).

The Development and Restoration Appeals

The Fire created a particularly serious financial situation in relation to the School's planned development programme. The new Recreation Centre was in the process of construction and a development appeal for the project, with a target of £500,000, was mid-way through its course. A decision was quickly taken to continue and complete this appeal in order that the Recreation Centre could be officially opened in September 1980. The Development Appeal for the Recreation Centre was finally concluded having raised a total of £517,000.

A Restoration Appeal was then launched in November 1980 with the large numbers of Old Bedfordians who had not been approached during the Recreation Centre campaign and new Parents of boys coming to the School.

Because of the wide-spread distribution of Old Bedfordians, the campaign has taken more time than its predecessor, although the results have been equally satisfactory. The final target for the Restoration Appeal has been set at £600,000 and the Appeal will be completed by the end of 1984.

In the School's hour of need, all its supporters have risen to the occasion quite superbly. Help from Old Bedfordians and Parents has been magnificent. There has also been most useful and solid assistance from a number of trusts, as well as from certain firms and companies.

The financial structure of both appeals has been based mainly on a range of covenants, or their cash equivalents, the amounts varying widely between pre-planned limits. The aim has been to achieve a gross average of £500 per head over seven years, and both appeals have comfortably exceeded this average.

The following are lists of those who have so far covenanted contributions towards both our Appeals. We have, in addition, received many generous single donations which have helped to swell the overall total and which will be recorded on a final list in due course.

To all our contributors we extend our most grateful thanks.

R. A. Franks

The Development and Restoration Appeals

Development Appeal

A. C. W. Abrahams
J. Adams
A.E.R. Limited
J. E. Alcock
H. W. Allardyce
A. J. Allen
R. M. Allen
A. R. Allen
A.P.E. – Allen Limited
T. C. Allerton
M. G. Amos
S. Anderson
W. Anderson
Dr R. H. Andrews
Anglia Equipment Supply
   Limited
Anglia Television Limited
J. D. V. Appleby
W. D. M. Archer
R. B. Attenborough
D. J. P. August
Mrs E. M. Austoni
N. J. B. Ayliffe

J. L. Baber
M. D. Baber
B. D. Bagot
A. R. Banks
R. L. Banks
Sidney C. Banks Ltd
Barclays Bank plc
A. E. T. Barcock
T. A. Barden
J. D. Bardner
Major J. S. Barker
A. M. Bark
K. D. Barker
M. E. Barlen
B. E. C. Barnett
R. Barnard
A. B. Barnes
C. B. Barratt
Mrs J. V. Bartle
A. H. P. Bartlett
M. K. Bass
D. G. Bates
M. D. Batho
Dr J. H. Baylis
W. Beazley
H. W. T. Beckett
T. J. Bedford
P. S. F. Belsham
A. Bell
D. L. Benham
R. Bennett
J. A. Berry
J. S. V. Bevan
Brigadier N. H. Birbeck
Mrs D. Birch
E. W. Billington
M. J. Birkert
S. Birkert
K. S. Biswell

N. Black
R. J. Blake
K. C. Blane
C. E. Blow
E. H. Blundell
S. Boddington
J. M. Boetius
G. M. Bolton
D. C. A. Bonnett
R. Bonnett
Canon D. H. Booth
Boots Charitable Trust
D. M. Boston
N. C. R. Boulting
D. M. Bowen
R. P. Bowles
H. B. Boys-Stones
W. J. Bradford
T. J. Bradshaw
D. A. C. Braggins
J. M. Brazier
N. C. Brewer
P. R. Brewin
M. L. Bridgeman
A. F. B. Bridges
G. R. Briggs
J. Briggs
P. D. Briggs
Dr R. A. Briggs
R. J. Briscoe
J. Bristow
British Schools &
   Universities
   Foundation Inc.
A. Brodie
D. N. Bromwich
A. W. Brown
D. W. Brown
R. Brown
R. J. Browning
M. P. Bryant
T. A. Bryant
M. O. C. Bulgin
C. S. Bunker
D. G. Bunker
G. B. Bunker
Surg.-Cdr. W. A. Burnett
   OBE
I. K. Burns
Henry Burt & Co. Ltd
A. G. Burton
J. W. Bushby
C. J. Bushell
B. Butters

G. E. Cacanas
S. J. E. Caen
Dr P. H. Calder
A. N. Caldwell
Dr E. Cameron
R. D. Campbell
I. Campbell-Gray
D. N. Capon
T. Capon
Dr J. L. H. Capper

H. R. Carlisle
P. R. Carlisle
J. K. Carlton
D. J. G. Carr
C. D. Carrington
G. A. Carpenter
J. S. Carter
G. Cartmell
J. B. Cartwright
R. L. Cartwright
Dr A. J. Cashman
J. H. Castle
P. Caswell
D. T. Caves
B. Cawley
P. W. Chamberlain
M. J. Champion
Dr M. Chapman
H. J. Charie
T. W. Charlton
I. Chichester-Miles
P. J. W. Chilton
R. E. H. Christie
I. Chrystal
H. M. Chung
S. J. Clear
E. F. Clark
Mr & Mrs J. Clark
Sqdn.-Ldr. K. P. Clarke
A. Clayton
Dr R. D. Clements
D. J. Clifton
P. R. Clint
J. R. G. Clover
A. M. Coates
T. E. Codd
I. C. Codrington
A. H. Coggins
L. Coladangelo
I. F. Colbrook
D. A. Coleman
R. J. Colling
L. D. Collins
P. W. Collins
D. H. Commins
Prof. C. J. Constable
M. J. Cook
J. Cooper
John Corby & Sons
   (Bedford) Ltd
P. V. Cope
Mrs E. E. Copperwheat
M. F. Copperwheat
K. W. Corden
H. Costello
P. R. Cowell
P. Cowell
P. N. Cox
Dr C. C. Crampton
Rear-Admiral R. W. D.
   Cook
Mrs J. R. Crawford
D. S. Crawford
E. F. Crawford-Smith
M. C. H. Cray

P. E. Creed
Mrs M. R. Crosland
H. C. E. Culverhouse
I. P. E. Culverhouse
Brigadier J. G. Cumberlege
C. C. Cumming
Mrs E. Cumming
P. S. Cummings
F. M. Cunningham
N. C. Curran
C. F. J. Curtis
C. J. Curtis
E. W. Curtis
Cutler Hammer Europa Ltd
I. W. Cutress

Lt.-Co. W. Daniels
C. R. D. Danby
M. C. Daniel
Davison & Co. (Barford) Ltd
A. G. Dart
J. G. Davies
Lt.-Col. M. J. K. Davies
P. J. Davies
G. Davis
I. G. Davis
R. M. Davis
R. S. Dawes
R. D. Dawson
J. K. N. Dawson
R. F. Dawson
D. G. Day
F. M. Day
A. F. Deards
M. Delaney
P. G. G. Dell
D. L. Dempster
Mrs J. U. Desroy
R. G. Dewe
R. Digby-Clarke
R. V. Dite
B. J. Dimmock
W. P. Donald
G. F. Downes
R. A. Downes
Dr M. F. Downey
A. G. Drayson
S. Duckworth
Duckworth & Kent Ltd
   (T. A. Waldock)
C. Duncan
H. G. Duncan
L. P. M. Duncan
C. B. Dudeney
J. M. Dudeney
D. M. Dunkeley
N. A. Dunne

B. C. E. Earle
C. O. S. Eales
Col. L. C. East
C. H. Eckert
A. T. Ednie
J. R. Edwards
C. W. R. Edwards

R. J. Elam
S. E. Eley
T. J. Elliott
G. J. Elliott
R. M. Elliott
C. Elphee
H. L. Ensor
V. S. Eltringham
A. D. Eraut
Lt.-Col. R. B. S. Eraut
D. A. W. Evans
A. W. Evans
G. T. Evans
W. Ewing

B. F. Falcongreen
S. C. Y. Farmbrough
J. D. Farnworth
G. R. D. Farr
E. J. Faucett
G. E. Felton
Dr J. Fenton
M. J. Ferdinando
R. F. Finch
B. Fielden
J. A. Fielding
Col. D. I. M. Finlayson
George Fischer Castings Ltd
R. B. C. Fitch
D. Flint
Mrs G. R. Flude
R. France
J. W. Francis
J. K. Frankish
G. P. Franks
R. A. Franks
A. R. Freeman
French Kier Holdings Ltd
Mrs E. S. Frost
M. Freyhan
Capt. A. T. Foster
A. F. Fox
Mrs S. W. Fozzard
D. Fu Chi Hung
A. R. Fuller
A. J. Fullerton

P. A. Gair
A. Gaishauser
R. G. Gale
F. N. W. Gamman
C. H. H. Garey
C. F. M. Garner
G. G. Garner
D. Garrett
L. F. Garrett
A. A. Garson
Wg.-Cdr. P. A. Garth
A. N. Gavin-Fuller
H. Gell
Geerings of Ashford Ltd
Dr D. M. George
Dr D. L. Georgala
Frederick Gibberd &
   Partners

A. J. G. Gibbs
Gibbs & Dandy Ltd
A. D. Gibbins
R. Gibby
D. Gibson
J. S. Gillett
Lord Godber of Willington
H. E. Godfrey
Major J. N. A. Goldsworthy
E. J. S. Gooch
G. H. Goodwin
R. E. Gordon
Dr A. A. Graham
Granada TV Rental Ltd
M. A. Grant
D. S. Gray
L. J. Green
A. N. Green
T. E. R. Greenaway
A. Greene
W. C. Greening
Dr B. V. I. Greenish
K. J. Greenwood
I. J. Greeves
Mrs V. R. Gregory
A. R. F. Griffin
D. J. Griffin
Mrs F. J. Gompertz
J. E. Grose-Hodge
Mrs S. Groves
Mrs J. Grundon
J. W. Gudgin

D. R. Habberfield-Bateman
J. I. H. Hadfield
A. Hakim
J. Hale
J. A. Hale
M. E. Hale
J. B. Hall
J. R. Hall
F. N. Hallifax
S. L. Halse
J. Halton
Dr B. H. Hamilton
B. A. Hammond
S. B. Handley
W. E. H. Handley
T. I. Handscombe
J. R. T. Hansel
P. B. Hansford
R. Harbinson
Mrs E. Harling
C. E. Hargreaves
R. J. Harper
M. D. Harper
G. O. Harries
C. J. Harris
R. R. Harris
F. H. Hartley
G. Hartwell
Haslemere Estates Ltd
Sqd.-Ldr. M. H. Hattan
M. J. Hawkes
J. R. Hawkins
Dr W. Hawkins
M. P. Hay

E. C. Haywood
M. P. Heckler
C. S. Hemsley
L. J. Heaton-Armstrong
H. R. Henley
J. C. Hendry
M. B. N. Henman
F. C. Henshaw
R. Herbinson
D. M. Herd
Lt.-Col. B. J. Herman
G. O. Herries
G. W. D. Heslett
N. D. Hewison
R. J. P. Hewison
J. M. Hext
D. T. Hicks
B. Higgs
Lt.-Col. D. R. Hildick-Smith
B. E. Hill
M. L. Hind
J. J. P. Hine
C. R. Hipwell
A. G. A Hodges
E. Hodson
N. F. Hodson
R. A. Hodgson
J. J. E Hodkinson
M. W. Holes
A. H. Holladay
J. P. Holland
V. J. Holley
D. R. Hopkins
R. D. R. Hopkinson
D. J. L. Hoppe
A. E. L. Horrocks
K. B. Horton
L. M. Hotchkies
E. A. S. Houfe
G. C. Houghton
P. T. Housden
C. A. E. C. Howard Ltd
Sir John Howard
J. A. Howard
P. Howard
C. J. Howe
S. Howe
T. L. Hughes
P. F. Hulance
C. J. Humphris
M. D. Hutchinson
Major J. C. R. Hyde

J. S. Inman
B. R. Inman
H. T. Inskip
J. G. Inskip
P. T. Inskip
Mr and Mrs D. E. Issitt
S. Ives

Dr P. G. Jackson
Mr and Mrs B. L. James
T. M. D. James
A. S. Japhet
Mrs P. E. Jenkins
Lt.-Col. A. R. Jesty

J. T. Jew
N. O. Jewers
Mrs M. Joiner
D. G. Johnson
G. A. C. Johnson
I. M. Johnson
D. B. Johnston
D. N. W. Johnstone
B. N. Jolles
B. R. Jones
C. I. M. Jones
J. S. Jones
T. Hamer Jones
P. W. Jones
M. A. Jose
A. I. Joseph
Lt.-Col. I. H. Joseph
L. A. W. Jordan
W. Jordan & Son
  (Biggleswade) Ltd
W. J. Jordan

A. Kavan
Mrs P. Kavanagh-Dowsett
R. L. Keech
Major J. A. Keer
A. J. Keiller
Dr M. J. F. Kelly
R. P. L. Kelham
W. A. G. Kendall
M. J. Kenningham
Lt.-Col. D. Kenwrick-Cox
C. J. R. Kettler
A. J. Keyworth
H. R. M. Killender
D. Kimber
D. King
B. R. Kingston
R. L. Kingston
General Sir Sidney Kirkman
D. A. Kitchiner
J. B. Knight
D. L. Knights
F. Knightly

E. J. Lacey
I. M. Laing
Rev. J. C. Laird
G. H. Lampe
R. F. Lander
W. Landmann
B. W. Lavis
Cdr. M. C. Lawder RN
The Hon. H. de B. Lawson-
  Johnston
H. H. Lawson
G. H. E. Layton
J. H. D. Lee
P. F. Lee
R. A. Lee
Major-General F. St. D. B.
  Lejeune
S. H. Leung
G. S. Lewis
K. C. Lewis
R. T. Lines
Col. R. R. Lindsay

Dr. B. Linganayagam
D. F. Linton
D. J. Litson
G. P. F. Little
R. Littlehailes
Lloyds Bank plc
J. F. Lobo
Dr B. Logan
C. A. Logan
London Brick Co. Ltd
Dr R. H. N. Long
P. A. Lousada
T. S. Luddington
Major J. R. Lumley
A. Lynes
N. Lyster-Binns

T. Machin
H. G. G. Mackay
H. T. G. Mackay
Mrs M. I. Mackechnie
Mrs W. Mackenzie
G. H. Maclean
G. J. R. MacLusky
B. H. MacNay & B. W.
  MacNay
L. V. McDonald
Sir Basil McFarlane
B. F. McGinty
M. A. McGrath
Mrs B. J. McGruer
D. J. McLeod
Mrs G. A. Makela
A. Malek
C. F. Mann
J. Manning
Major P. B. Manson
Mrs M. Markham
D. A. Marks
R. L. Marshall
H. Martell
Dr E. E. J. Martin
J. T. Martin
N. C. G. Martin
Mrs & Mrs P. J. Martin
R. K. Master
J. Matthews
J. P. Maudlin
T. J. Maynard
V. A. D. Mayes
B. A. F. Maynard
P. A. Mayo
P. T. Mead
Col. W. P. Meldrum
J. S. Mence
E. A. Mercer
Air.-Cdr. H. L. Messiter
N. P. Michie
Midland Bank plc
J. Millard
J. S. Millar
A. S. Miller
G. Millman
Dr A. M. D. Mitchell
W. Moeller
Dr D. C. Monks
P. F. Morgan

M. W. L. Morris MP
D. B. Moore
Lt.-Col. D. J. R. Moore
G. B. Moore
J. Moore
Lt.-Col. J. P. Moore
M. E. Moore
P. H. Moore
E. Morley
S. G. Muir
S. K. Mukherjee
D. F. Munday
Dr S. B. Murnal
A. Murray
Mrs J. A. T. Murray
R. Murthy
L. Myers

National Westminster Bank
  plc
G. A. Neale MP
A. M. Neely
Mr and Mrs H. B. Neely
A. Nethersole
P. B. Nickels
G. W. Nichols
A. D. Nightall
A. C. Noble
Mr and Mrs D. P. Norris
J. W. Northern
R. R. Nudd

Lt.-Col. C. L. Oakley
M. H. D. O'Callaghan
B. S. O'Dell
R. C. O'Dell
J. C. C. Oliver
P. Orchart
A. H. Ormerod
G. A. Orr
R. S. Osborne
N. I. E. Ostram
Dr D. R. Owen

C. W. Pagan
H. P. Pagett
G. H. J. Paisey
B. M. Palmer
F. S. Panton
Dr N. M. Panton
B. M. B. Pape
G. M. Parish
Dr W. E. Parish
Sir Peter Parker
Col. R. F. Parker
R. J. Parnwell
Peacocks
A. J. Peacock
E. D. M. Peacock
G. Peck
M. W. Pebody
W. G. Pebody
R. H. Pell
N. S. Pemberton
J. N. Penfare
N. W. Penn
L. H. E. Perks

D. A. Pettifar
P. Pettigrew
V. F. Phillips
Dr J. S. Phillpotts
J. N. Phillpotts
A. H. Phypers
D. W. M. Pinkney
Rev. W. E. Pitt
G. Place
T. J. Prescott-Brann
Sqd.-Ldr. R. C. Pocock
L. W. Pollard
J. C. Pope
J. H. Porter
Dr P. Porter
H. D. Praat
T. H. Preston
M. J. Prosser
M. C. Purbrick
J. Pybus
W. M. Pybus
Dr P. O. Pyle

T. A. Quinn

F. R. Radice
Mrs S. Rainbow
P. Ralphs
G. Ramm
A. H. Randall
J. A. Randall
M. M. Ratcliffe
M. J. Rawlinson
Dr P. Rawson
S. J. Rayner
Dr P. C. Rea
R. T. A. Read
Dr S. Reddy
A. R. Reed
C. N. Reed
Sqd.-Ldr. R. J. Rees
Dr D. J. Rhodes
C. Richardson
J. G. C. Richardson
J. D. Richardson
J. H. Richardson
B. Richman
Lt. Col. L. T. G. Ricketts
P. J. Ricketts
Dr W. D. Riding
P. W. T. Riley
T. B. H. Riley
C. W. Rippon
J. Rixon
P. W. Roach
C. E. S. Robbs
B. P. Roberts
T. A. Roberts
G. H. Robertson
W. W. S. Robertson
G. A. Robinson
P. F. Robinson
B. C. J. Rogers
D. B. Rogers
D. P. Rogers
D. J. Rogers
N. A. Rogers

Mrs Ruth Rogers
E. T. Rollinson
K. G. Rose
L. J. Rose
J. S. Rosser
D. W. Russell

A. R. J. Sabey
R. A. Sales
J. E. Sansome
Mrs S. Sarsfield
G. P. C. Saunders
J. H. T. Saunderson
A. J. J. Saunderson
A. N. Savage
J. Sawford
Dr P. R. McH. Scales
H. F. Schwimmer
H. Schurink
W. A. Scott
D. Scrutton
B. M. Scully
A. J. Seager-Smith
C. R. Seaman
M. D. Sear
P. J. Searby
A. M. Semark
J. B. Sewter
J. R. Sexton
S.F.I.A. Educational Trust
J. M. Sharman
J. R. & R. F. C. Sharman
Major M. M. Sharman
J. A. Shelton
R. W. Sheppard
E. H. Shrive
V. L. A. Simmons
D. A. Simpson
J. B. Simpson
T. Simpson
B. Skinner
W. R. Slater
A. J. Smeath
Dr. A. D. Smith
A. W. Smith
B. E. Smith
C. P. Smith
E. J. Smith
J. A. Smith
Lt.-Col. J. B. Smith
N. B. Smith
P. J. Smith
Mrs T. J. Smith
P. A. South
Wing-Cdr. R. F. Sowerby
P. R. Sowman
F. W. D. Spinks
E. T. L. Spratt
E. G. Squires
K. Srinivasan
Mr & Mrs J. N. Staddon
R. Stainer
B. T. L. Stairs
Mrs N. E. Starkie
T. O. M. Starkie
A. H. Stearns
S. B. Stearns

Dr R. W. Steed
Lt.-Col. D. R. Stenhouse
J. M. G. Stenson
A. R. Stevens
Dr J. L. Stevenson
R. W. L. Stidolph
Lt.-Col. P. W. Stock
R. Stone
R. G. Stowe
A. M. Stonebanks
Lt.-Col. J. A. Stewart
Dr D. A. Stringer
D. M. D. Swaney
P. G. Swift

B. T. Tam
P. Tanner
T. A. Tansley
G. A. Tarrant
Dr F. R. Taylor
Mrs M. A. Taylor
M. B. Taylor
H. A. N. Tebbs
Texas Instruments Ltd
S. G. Thair
Dr K. Thiagarajah
M. R. Thody
A. J. Thomas
L. I. Thomas
M. S. Thompson
A. M. Thorp
D. A. Timperley
Mrs J. M. V. Tinworth
Tobler Suchard Limited
W. Tolson
B. Toyn
A. B. Trask
G. F. Trunchion
D. H. Turner
M. J. B. Turner
T. R. Twigg

P. J. Upstone

B. G. Wakerly
Mrs J. D. Walbank
G. W. Walker
Dr H. J. Wallace
I. H. Walrond
Dr D. Warbrick-Smith
Dr H. A. Warbrick-Smith
Bernard Ward (Bedford) Ltd
Capt. W. G. Ward-Smith
Ward White Group Ltd
R. E. Warlow
A. R. Van C. Warrington
R. N. Watling
D. G. Watson
J. S. Watson
Mrs M. J. Watson
Major J. P. Weatherall
R. Wedgbury
Charles Wells Ltd
D. Welti
G. T. West
G. H. F. Wheatley
T. M. Wheatley

Capt. P. G. R. Whitaker
K. Whittaker
Whitbread & Co. Ltd
Major Simon Whitbread JP
Dr D. J. White
K. L. White
J. K. White
Dr R. J. White
Dr W. T. White
Major S. B. Whitmore
J. H. Willey
S. A. Wilkinson
D. E. Williams
J. F. Williams
Dr T. P. Williams
W. G. Williams
J. B. A. Willis
J. T. Willmin
John Willmott (Bedford) Ltd
F. R. F. Wilson
A. J. Wimbary
R. D. Wisbey
I. D. Wood
T. C. H. Woodbridge
L. M. Woodley
B. G. Woodrow
A. W. T. Woodward
J. M. Woolfenden
J. R. Worboys
B. Worsdall
M. N. Worthington
Dr D. G. Wray
W. J. Wray
Mr & Mrs G. T. Wrench
Mrs P. J. Wrigley
M. G. Wright
Mrs Joan L. Wroe
A. L. Wylie
Lt.-Col. C. St. A. Wylie
G. B. Wylie

Dr G. Yerbury
F. E. Yorke
D. T. Young
P. A. Young
I. Yovichic

Dr M. Zaki

Restoration Appeal

T. W. J. Abbiss
P. B. Adams
Dr R. Addleman
J. Addison
R. P. Agnew
H. K. Aitken
S. R. Albright
H. G. Aldous
E. C. L. Allen
Alliance Building Society
Allied Breweries Ltd
V. G. de M. Ambrose
A. R. Anderson
R. L. Anderson
Lt.-Col. D. H. Andrews
Anglia Building Society

The Annenberg Fund Inc.
J. V. C. Anthony
E. M. Aplin
M. J. Appleton
Arbuthnot Latham & Co.
  Ltd
R. T. Argent
M. J. B. Arthur
P. W. Armstrong
R. P. E. Ascham
M. F. Ashley
R. A. Ashworth
L. Atkins
P. G. Astles
The Rev. R. C. Atkinson
The Augustine Trust
C. J. Austin
Mrs J. J. Ayers

L. C. Bailey
Sqn.-Ldr. K. W. Baldock
N. S. Baldwin
B. H. A. Ball
R. E. Bance
Dr M. J. Banham
Sidney C. Banks Ltd
R. R. St. J. Barkshire
C. I. Barber
Flt-Lt. R. E. Barber
Barclays Bank plc
Baring Brothers Ltd
A. T. Barker
B. K. Barker
D. A. Barker
Mr Justice Dennis Barker
P. M. Barman
C. F. Barnes
P. M. Barnes
R. S. Barnes
M. J. Barnett
S. Baskerville
P. J. M. Battle
J. H. Bayfield (Tobler
  Suchard Ltd)
R. M. Bayfield
G. E. Bayley
G. G. Bayley
B-Ch. Charitable Trust
A. V. V. Beaty
Bedford Building Society
T. J. Bedford
Bedford School Common
  Room
Bedford School Mothers
  Guild
C. C. Beeley
J. Bennett
P. J. Bentley
T. A. Berrecloth
A. E. Berrisford
S. T. Betts
A. V. Bevan
Dr E. V. Bevan
O. V. Bevan
G. Billing
The Rev. H. Birch
P. J. Bird

Bishop of Bedford – D. J.
  Farnborough
Mrs P. B. Blackburn
Group-Capt. A. B. Blackley
P. M. Blake
P. R. Bligh
Dr D. L. Blundstone
O. S. P. Blunt
Mrs J. Bolton
P. J. Bond
Dr J. W. Bone
A. R. Bosworth
Mrs J. Booth
Viscount Boyd of Merton
B. J. Bower
Mrs H. Bradfield
C. M. N. Bradford
D. A. C. Braggins
M. T. Branson
British Schools &
  Universities Foundation
D. J. Britton
J. R. Brown
J. S. Brown
P. Brown
Dr P. P. Brown
Mrs and Mrs R. Brown
The Rev. W. M. Brown
Brown Boveri Kent Ltd
G. E. Brough
A. G. Buchanan
I. Buchanan
R. A. Buckle
D. Y. Buckingham
L. H. Bucknell
A. E. Bugle
M. H. Bull
J. Bullivant
J. H. Bulmer
J. Bunyan
L. Butler
A. G. Butters
C. P. Burns
J. P. U. Burr
G. G. Burton
J. R. Bygraves
A. E. Bygraves

Mrs C. G. Cain
R. W. S. Cain
Dr W. H. Callander
M. Cameron
R. Cammack
B. D. Cantle
H. Castenskiold
J. I. Castle
Capel Charitable Trust
D. A. Capon
Mr and Mrs R. P. Carpenter
C. J. Carroll
Mrs P. N. Carroll
W. B. Carruthers
J. S. Carter
P. M. M. Cashman
A. J. Cave

C. P. Cawthorn
Lt.-Col. G. Chadwick
C. L. Chan
H. C. Chan
W. C. L. Chan
A. H. Chapman
C. S. Chapman
D. A. Chapman
M. D. P. Chapman
P. J. Chapman
Sqn. Ldr. B. J. Charlton
A. E. Charnley
P. Charsley
Charterhouse Group Ltd
C. P. G. Chavasse
P. A. G. Chavasse
R. A. Cheke
H. W. Cheng
P. J. Chignell
R. Chignell
J. B. Childs
Dr T. A. Choudhry
P. R. Churchyard
H. A. Clark
S. S. Clark
R. P. Clarke
T. J. Clarkson
The Rev. B. H. Y. Claxton
R. B. Claxton
J. E. Clayton
M. Clemens
Mrs E. M. Codrington
W. J. Coker
J. R. Cole
Wg.-Cdr. R. A. Collinge
P. L. Colvin
D. Coley
G. A. Colley
Dr S. D. W. Collier
Capt. C. A. M. Comins
T. Compton
P. D. Conniff
E. J. Coode
Dr J. Cook
S. H. Cook
A. M. Cooper
R. Cope
R. M. A. Corkery
Mrs B. A. Corley
Dr W. R. Cotton
C. A. Courtenay
Major J. D. P. Cowell
R. J. Cowell
The Lord Cozens-Hardy
  Glaven/Nimrod Trust
G. M. Crawford-Smith
E. H. R. Craig
G. A. W. Crane
J. L. B. Crane
P. J. Crankshaw
Mrs J. U. Croot
J. G. P. Crowden
Culra Charitable Trust
S. Culliford
K. R. Cumberland
J. D. Cunningham
Dr J. L. Currie

N. S. M. Cuthbert

J. H. Daker
J. A. Dakin
W. A. Dalgety
Mrs Dalgety
W. R. Dalzell
J. P. Daniell
G. S. Darlow
D. M. Darlow
Dr J. M. Darlow
Surg.-Cdr. H. M. Darlow
G. K. Davey
P. S. J. Davidson
Mr & Mrs G. Davidson
G. T. Davies
N. T. Davies
E. A. Davies
R. H. Davies
M. B. Davies-Jenkins
T. H. Davies
Wg.-Cdr. P. T. Davies
A. H. Davis
G. T. Dawson
M. Day
W. R. Day
J. Day
D.C.W. Trust
M. J. Deacon
E. M. Deacon
C. J. Deane
J. S. Deane
A. F. Deards
R. De Morgan
P. E. Devenish
I. F. Devereux
P. C. T. Dickins
Mrs E. L. L. Donovan
J. G. Doubleday
G. C. Draughn
A. W. Duckett
P. B. Dudeney
N. E. Dudley
J. L. Duffus
P. H. H. L. Duke
A. C. Dutton
F. J. W. Dyer
The Dyers' Company
R. W. G. Dynes

J. A. Eastcott
I. D. Eakins
B. M. Eckersley-Hope
Dr B. W. Eden
H. T. Edgecombe
M. E. Edgerton
J. C. Edwards
G. Elliott
The Elvetham Charitable
  Trust
Dr W. R. Emmerson
S. J. Endersby
C. J. England
P. E. Erskine-Murray
D. A. Escott
P. A. R. Evans

K. Everest
N. K. Eyre

Major P. C. Farmer-Wright
Rear-Admiral K. H. Farnhill
S. F. Faulkner
R. Faulkner
Dr B. M. Feilden
C. R. Feilden
Dr G. B. R. Feilden
P. W. Fenn
D. M. Fenning
A. Fields
N. H. Filskow
A. Findlay
Findlay Publications Limited
D. L. Fish
A. C. Fitt
Dr O. P. FitzGerald-Finch
Maj.-Gen. H. R. B. Foote
R. Forrest-Hall
G. L. Forster
J. Foster
S. A. Fleming
D. S. Fletcher
R. Framp
A. Franklin-Jones
G. A. Franklin
Sir Bruce Fraser
P. G. Fraser
I. F. H. French
Friends of Bedford School
Dr I. C. Fuller
R. K. Furbank
D. V. Furlong

R. T. Gallie
Mrs A. M. Gamble
A. A. Game
A. T. Gardner
A. C. Gardner
S. C. Gardner
B. Gardner-Hopkins
C. J. W. Garrett
D. A. B. Garton-Sprenger
M. B. Garton-Sprenger
I. B. Gaskell
A. C. Gaskell
B. K. Geary
R. R. Geering
R. E. Geeves
Sqdr.-Ldr. C. M. Gerig
J. K. Gerrard
Simon Gibson Charitable
  Trust
The Rev. H. W. Giddings
P. G. Gilbourne-Stenson
Wg.-Cdr. L. M. Gilchrist
Dr J. N. G. Gilchrist
M. M. Gilmore
P. J. Gillett
W. H. Gilliland
P. S. Gleave
R. T. Godber
Mrs A. M. Godden
Mrs G. D. Goodacre
R. J. Goodes

B. D. E. Goodman
R. E. Gordon
J. F. Goudge
M. R. Green
The Rev. B. Grainger
D. G. Grant
S. C. Gray
N. A. Green
Dr J. M. W. Grieve
E. E. Griffiths
The Grocers Company
N. Grove
N. S. Grundon
Arthur Guinness Son &
  Company Ltd
C. A. Gude
B. P. Gwynn

P. J. Hadfield
R. P. Hairsine
M. F. Hall
H. J. Hall
J. M. Hammond
L. E. Handford
M. J. C. Harbour
Dr P. B. Hardwick
Dr H. N. W. Harley
Cdr. C. H. A. Harper
W. D. Harper
W. S. Harpur
R. M. Harral
Dr J. A. Harrington
D. W. Harris
A. D. Harrison
W. I. Harrisson
D. J. Harris
F. J. Harris
C. P. Haslam
R. P. Hart
Chief Superintendent D. J.
  Hart
B. Hartley
P. J. Hartley
D. C. Hartley
G. R. Harvey
A. W. Havil
G. R. Hawkes
Dr S. S. Hawkins
D. S. Hayden
Haymarket Charitable Trust
C. C. S. Hayne
H. D. Hayne
C. W. J. H. Hayter
J. D. S. Hay
Headley Foundation
D. R. L. Heald
M. G. Hearth
T. V. Heathcote-Hacker
R. E. J. Hemmings
A. H. Hemsley
G. R. Henderson
W. H. Henderson
T. H. S. Henderson
C. A. M. Henniker
Rev. M. D. A. Hepworth
M. R. M. Herbert
R. E. Hersee

G. W. Henshilwood
R. G. Henton
J. H. Hewlett
Lt.-Col. P. H. B. Hickman
J. L. Hicks
Lt.-Col. D. R. Hildick-Smith
R. G. C. Hill
D. E. Hill
A. V. Hills
M. G. Hilson
The Lady Hind Trust
M. D. Hine
H. H. Hobbs
J. A. Hobley
Dr J. Hockey
A. G. A. Hodges
Lt-Col. Q. D. T. Hogg
R. Holbrook
G. W. L. Holden
A. Holden
L. Holland
A. H. D. Holland
B. F. Holman
W. N. Holme
R. A. Holton
P. D. Hoole
Mr & Mrs J. D. & R. E.
  Hopkins
H. L. Hoppe
Major P. B. L. Hoppe
A. H. L. Hoppe
P. E. Hoskins
J. A. Houghton
Sir John Howard
R. G. Howard
C. N. Howarth
D. G. Hudson
M. Hughes
Vice Admiral Sir Charles
  Hughes-Hallett
D. B. T. Hughes
Rev. J. A. L. Hulbert
D. N. Hunt
P. Husband
R. N. Hutchins

L. E. Imroth
Mr & Mrs R. Ingle
D. Ingledow
D. J. G. Ireson
C. T. Izzard

A. Y. A. Jiwaji
D. R. Johnson
J. G. Johnson
R. E. Johnson
The Reverend Dr W. Barr
  Johnston
M. Joll
T. W. Jolly
D. C. Jones
Dr M. E. Jones
E. A. W. Jones
C. N. W. Jones
D. A. Jones
F. G. Jones
A. N. Jordan

J. R. Joyce
Dr P. J. Joy
R. S. Juffs

E. W. T. Kaye
S. F. Kellett
J. A. Kempton
P. A. Kennedy
Dr A. Kennedy
T. D. Kent-Jones
H. G. Kettleborough
R. L. J. Kilby
M. P. King
Kleinwort-Benson Limited
Dr S. Kownacki
Dr R. U. F. Kynaston

M. D. Lacey
Laing & Cruickshank
  (R. Dawes)
Laings Charitable Trust
Mr & Mrs L. P. K. Lam
Dr G. D. Langham
J. M. Langham
J. G. Langley
M. J. M. Lawley
P. Laws
P. C. F. Lawson
J. H. Lawson
G. H. E. Layton
A. S. Leaver
A. R. Lee
V. F. D. Lee
J. A. Ledger
D. A. Ledsom
C. J. E. Legg
A. Leigh
S. G. Lersch
I. R. Letham
Letraset Limited
  (J. D. Bardner)
S. H. Leung
J. W. S. Lewis
G. Lewis
J. Lewis
Major-General H. M.
  Liardet
D. W. Lilley
R. M. S. Lincoln
Dr D. R. M. Lindsell
D. Lines
T. C. Liu
B. M. Lippmann
The Corporation and
  Members of Lloyd's &
  Lloyd's Brokers
G. F. G. Lloyd
Lt.-Col. J. Lloyd-Rees
A. C. Lochhead
London Law Trust
A. N. O. Long
J. B. Low
Dr R. L. Lucas
C. Lumley-Wood
I. M. Lyle
Mrs B. Lynn
R. Lynn

G. J. H. Mackie
Dr G. I. Mair
M. H. Mallet
R. L. Mallett
H. D. Malley
Major D. R. Mantell
V. P. Manze
J. P. Marchant
Mrs S. B. Mardlin
R. W. Maris
T. Markham
P. S. Marks
Marks & Spencer plc
O. J. Marlow
M. J. Marriott
D. N. Marr
B. J. Marsh
B. L. Marsh
I. Marshall
Marshall of Cambridge
  (Engineering) Ltd
A. P. W. Martin
P. M. Martin
H. D. G. Martindale
Capt. C. Marton
G. T. Martyn
F. G. Mathers Limited
K. W. May
Prof. A. D. May
Col. P. A. Mayer
G. S. O. Mayne
A. T. Mead
Sqdn. Ldr. D. M. Meadwell
D. J. Measham
M. D. M. Memorial Trust
Mercantile House Holdings
  Limited
The Mercers' Company
M. A. Metcalfe
Dr G. C. Metcalfe
A. Michell
E. B. Middleton
J. N. Miers
M. H. Miller and A. A.
  Miller
R. B. Miller
D. N. Miller
A. C. Miller
L. W. Miller
E. A. Millson
J. P. Mills
A. I. Mills
T. G. Mitcham
K. Mok
R. H. Moll
G. M. Moll
E. C. Moll
Dr M. Momen
C. P. Money
F. B. Moulang
R. T. Montgomery
G. Montgomery-Smith
Dr R. H. Morris
N. B. Morrison
Miss S. M. Morse
M. J. H. Mortimer
J. R. M. Mortimer

A. R. Morton
Lt.-Col. I. D. Moss
M. E. Mulgrue-Green
A. G. Munro
Dr W. P. Munro
W. R. Murdie
T. Murphy
R. J. Murphy
J. O. Murray-Clarke
C. R. Murray
S. Murray
M. J. Must
J. M. Mutch

Judge A. C. MacDonald
Macdonald-Buchanan Trust
Judge J. R. Macgregor
I. MacKechnie
J. F. MacLeod
Dr T. MacLennan
C. MacMillan
C. A. S. McAra
A. McCallum
Wg.-Cdr. B. C. McCandless
Mrs S. McCarthy
Major P. S. McCullagh
Dr P. McMullen

National Westminster Bank
  plc
D. J. H. Neate
G. M. Nell
P. G. Nelson
R. L. K. Nice
M. R. Nevard
W. O. Newcomb
Cap. H. M. Newcombe RN
A. M. Newell
NSS Newsagents Limited
Dr J. Newton
C. F. M. Norman
J. J. Normoyle
Lt.-Col. F. G. Norton-Fagge
S. J. Novis

L. G. L. Oakley
Lt.-Col. D. R. Oakley-Hill
P. R. M. Oaten
Major B. M. O'Bree
A. E. O'Broin
Old Bedfordian Lodge
Brig. E. N. Oldrey
D. N. Oliver
Lt.-Col. C. L. Ommanney
Ormonde Charitable Trust
A. W. O'Rourke
W. R. S. Osborne
Dr G. A. Oswald
Dr J. R. Owen

J. D. Pagan
R. G. Palmer
S. A. L. Panton
G. E. Parker
M. Parker
Col. A. S. Parkin
M. N. Park

D. M. Park
M. J. Park
J. C. & R. G. Parrish
W. A. Parrish
A. J. Parry
D. C. S. Parsons
Mrs E. J. Paton
A. I. Paterson
Lt.-Col. D. J. Patrickson
L. G. Patterson
M. H. Pattison
Dr P. R. Payne
A. J. Peacock
J. M. Peacock
R. A. Peacock
Capt. J. H. T. Perera
J. O. N. Perkins
J. R. Pettley
P.F. Charitable Trust
Phoenix Assurance Company
  Limited
Phillips & Drew
J. W. Phillips
H. L. Phillips
G. H. Pickerell
R. Pickett
R. S. P. Pinguet
Lt.-Gen. Sir William Pike
Marshal of the Royal Air
  Force Sir Thomas Pike
M. E. Pile
H. G. L. Pitt
C. J. Planting
R. J. Pleasants
Grp.-Capt. P. V. Pledger
G. Pleuger
R. P. Plumpton
J. B. Plumptre
The Rev. R. A. Plumptre
B. J. Pollard
T. D. Pope
A. B. Porter
G. O'B. Power
H. D. Praat
Mrs M. A. Prain
J..Prentice
P. W. Previte
W. R. Price
G. D. Price
Mrs C. A. Price
J. R. Probert-Jones
P. S. Putnam
J. C. Putterill

R. W. Quartermaine

Wg.-Cdr. A. E. Radnor
The Rev. W. R. Railston-
  Brown
A. T. Rainbow
J. T. Rainbow
Dr R. M. M. Rajapakse
P. L. Rallison
N. R. Ramsay
Lt.-Col. A. M. St. L.
  Ramsay-Murray
The Rank Foundation

R. J. Rawlings
A. W. Read-Price
The Rev. D. J. Rees
Mrs W. A. Reid
G. W. Rich
S. J. Richardson
R. E. Richardson
Prof. R. W. Rideout
A. J. Ridler
Dr D. B. Rimmer
P. W. Ripley
W. Ripley
Prof. J. P. C. Roach
A. D. Robbins
M. J. Roberts
C. J. L. Roberts
M. Roberts
R. H. Robertson
F. A. Robinson
L. J. Robson
Sqn.-Ldr. M. Rodgers
Major J. W. Roe
C. R. Rogers
C. G. Rogers
R. A. B. Rogers
Dr J. H. Rogers
Mrs S. J. Rogers
E. T. Rollinson
Dr P. G. Rose
D. A. Roseveare
B. T. Ross
J. G. Ross
Rothman's International
Limited
J. H. Rowe
P. Rowland
D. G. Rowley
Royal Insurance Company
N. S. Rudge
J. W. E. Rumboll
A. L. Runeckles
K. P. W. Rutgers
D. Rutherford
P. A. Ruysen

E. M. Samy
J. V. S. Sahai
R. J. Salisbury
Major H. W. Sargeant
J. W. P. Sargent
Dr J. H. B. Saunders
R. J. Saunders
Mrs J. S. Savva
J. E. Saxby
R. A. Scott
Lt. M. S. Scrivens RN
Col. P. W. G. Seabrook
F. H. Seale
Major R. A. M. Seeger
H. P. Shallard
I. A. T. Shannon
J. R. Sharman &
 R. F. C. Sharman
J. Sharples
F. Shaw
M. H. Shaw
S. E. Shelton

Lt.-Col. P. O. B. Sherwood
T. J. Sherwood
R. P. F. Shorten
T. H. Shrive
C. W. Shuter
S.F.I.A. Educational Trust
D. S. Skinner
J. C. Skinner
Siebe Gorman Holding
Limited
D. C. Simco
Mrs M. I. Sizen
B. J. Sinfield
P. Singh
Mrs R. Silcock
Dr. M. J. Simmonds
R. G. Simmonds
G. W. Simmons
G. Simmons
Mr & Mrs A. J. Simpson
Mr & Mrs B. H. Simms
Slough Estates Limited
G. S. Smalls
B. R. Smedley
R. W. J. Smith
D. H. Smith
J. L. Smith
L. M. Smith
P. D. H. Smith
P. W. Smith
C. D. C. H. Smith
R. I. Smith
G. H. Smith
C. A. H. Smith
J. Smith
D. J. Smith
K. E. C. Sorenson
D. F. Soul
F. H. Sowman
D. R. Sowman
T. G. Sparrow
S. J. Speeks
D. G. & Mrs M. J. Spencer
J. A. Spooner
M. J. P. Srawley
M. P. Stambach
P. J. Stanley
G. Stanley
P. F. Stanton
Dr W. D. Steel
N. G. Steel
C. J. Steen
Rev. W. D. Stenhouse
L. C. Stephens
R. V. Stevens
K. A. A. Stevenson
M. F. Stevenson
H. M. B. Stewart
R. D. Stewart
P. Steyn
S. B. Strickland
M. I. Stillwell
Sqdn.-Ldr. R. D. Stone
V. G. Straubs
D. R. Stratford
A. G. Stroud
D. W. Stubbs

P. J. Stubbs
D. F. Studman
H. W. Suen
Major W. Surtees
D. M. Sutcliffe
J. R. Suter
W. E. Suter
R. J. H. Swainson
R. R. S. Swan
Dr A. J. Swannell
Mrs I. Swift
S. J. Symonds
J. C. T. Symons

B. S. Tansley
J. C. H. Tate
G. B. Tatman
The Rev. M. F. Tayler
The Marquess of Tavistock
J. D. Taylor
E. F. Taylor & Company
Limited
G. M. Taylor
D. W. Tear
Flt.-Lt. I. R. Tench
S. G. N. Terney
G. H. Thatcher
F. C. Thody
P. T. Thompson
R. W. Timson
P. A. Tinkler
C. H. Titley
C. K. Y. Tjon
N. A. Tobin
D. F. Tomlinson
W. H. Towle
Town & Country Building
Society
P. I. Tresham
J. Trewavas
Group-Captain O. J.
Truelove
M. J. Tuckett
A. W. Turnbull
Lt.-Col. P. E. X. Turnbull
C. G. S. Turner
J. O. Turner
R. Turner
R. H. Twinch

R. C. Uden
Unigate Limited
Unilever Limited
University Arms Hotel
T. J. Ullman

P. J. Valentine
C. D. Vane-Percy
J. N. P. Vann
J. Vaughan
J. V. J. Vivian
D. Von Winterfeldt
R. B. Waller
K. Waddington
P. J. Waite
S. C. Waldecker
Mrs M. C. Walker

T. J. Walker
B. J. Wilkins
G. E. Walmsley
P. S. Walsh
W. J. R. Warren
R. I. C. H. Warren
M. Warren-Smith
Dr M. P. Wasse
R. T. Waterfield
P. B. Watermeyer
W. Watson Buiders Ltd
J. Watson
M. S. Watson
Mrs S. L. Watts
R. V. Welborn
C. M. Wells
D. F. Wells
The Welton Foundation
C. H. S. West
C. West
P. R. Whaley
G. Wheatley
Whitbread & Company
Limited
Simon Whitbread Charitable
Trust
W. H. Whitbread Charity
Fund
Major D. B. White
J. White
S. H. T. White
M. S. White
Colonel P. Wildman
Mrs J. M. Wilks
J. Wilson
G. F. Wilson
N. P. Wilson
D. R. Wilson
Lt.-Col. F. A. C. Williams
G. A. Williams
G. B. Williams
R. A. D. Williams
S. E. Williams, G. F.
Williams and I. R.
Williams
G. E. Williams
A. K. H. Williams
D. R. Williams
T. W. Williams
G. N. Wills
A. J. Wiltshire
A. C. W. Wimbush
W. A. D. Windham
N. R. Winterburn
D. G. Winter
J. H. Winton
C. P. Wolf
N. Womersley
P. R. O. Wood
Gp.-Capt. A. A. G.
Woodford
J. R. Woodford
M. M. Woodford
E. A. Woodger
P. T. Woods
D. E. Woods
W. J. Woodward

E. Wootton
W. P. Worth
L. A. Wright
H. L. Wright
P. N. E. Wright
W. D. Wright
R. R. Wyatt
R. B. Wylie
Colonel K. N. Wylie

Mrs J. M. Yates
A. G. B. Young
H. V. M. Young
J. D. C. Young
A. H. C. Young

Members of Staff Easter Term 1979

T. C. Allerton
Miss S. J. Allison
E. J. Amos
D. A. Armstrong
Miss A. S. Awdry
D. C. Bach
K. Baddeley
M. E. Barlen
A. E. Barlow
P. Boorman
P. G. Bossom
N. C. R. Boulting
P. C. Braggins
P. D. Briggs
D. H. Bullock
D. E. Butler
R. G. Caple
Mrs C. F. Carter
M. J. Carter
P. B. Churcher
Mrs W. A. Clark
L. H. Coblenz
P. J. Coggins
P. E. Coles
J. I. Cooper
B. J. Cotton
N. J. Cox

R. A. Craddock
B. W. Crossman
Dr A. J. Crowe
Mrs C. N. L. Crowe
J. H. Davidson
Rev. P. R. Davies
R. A. Eadie
T. J. Elliott
R. J. Essam
Miss S. U. J. Findlay
A. C. Fitt
F. M. Fletcher
G. M. K. Fletcher
J. R. Fletcher
J. L. Foulkes
H. D. Galbraith
Dr A. D. Gazard
R. F. Goodacre
S. F. Graveston
G. Grimshaw
R. C. Hadley
K. S. Hall
P. Harrison
G. S. Haynes
Rev. M. D. A. Hepworth
W. Hill
I. F. Howe

D. B. T. Hughes
Miss S. M. Hughes
Mrs J. R. Hull
D. W. Jarrett
C. G. G. Jeavons
O. D. Ladd
Rev. J. C. Laird
Mrs B. Lowe-Ponsford
T. J. Machin
H. F. McKendrick
P. M. MacMahon
W. E. Makin
Maj. D. R. Mantell
R. A. Middleditch
A. R. Millett
D. F. Moore
M. J. Norris
J. B. D. Osborne
J. C. Osman
K. Parkinson
S. J. Partridge
H. J. W. Pidoux
G. F. W. Pleuger
D. C. Poulton
Mrs C. Powell
P. F. Ramage
Mrs I. J. E. Rawet

M. J. Rawlinson
J. S. Rees
Mrs F. M. W. Rhodes
T. A. Riseborough
R. P. Robinson
I. J. Rosser
A. S. Rothon
D. K. Sewell
Mrs N. B. Smith
M. P. Stambach
F. J. Steele
C. J. Steen
D. P. C. Stileman
J. Sylvester
A. M. Thorp
R. D. Tomlinson
I. R. H. Tovey
M. R. Vogel
R. D. Warren
D. N. Whatley
P. S. Wiser
P. R. O. Wood
B. Worsdall
P. A. Young

Names of boys by forms Easter Term 1979

VI/LVI

Ashpole R. N.
Robertson J. S.
Briggs S. W. J.
Frankish J. A.
Kenningham D. M.
Laird A. J. W.
Milne G. A.
Palmer R. W. D.
Pyle J. H.
Scrutton R. A.
Smith R. C.

HVIa

Ayliffe J. J. B.
Burch M. W. J.
J Evans C. D.

Stokes R. D.
West D. G.
Willis T. J.

HVIb

Benham A. D.
Clifton J. M.
Constable C. A. J.
Crampton R. M.
Harries G. M.
McKendrick I. A.
Nutt S. A.
Smalls N. G.

HVIc

Archer R. W.
Heathcote T. P. L.

Housden C. P. T.
Lebl D. N.
Rückeis T. J. J.
Smith R. T.

FVI/LFVIa

Ross Macdonald R. A.
Staddon J. A.
Belsham P. I.
Neumann K.
Bailey M. B.
Clayton S. P.
Flanagan P. A.
Heckler M. P.
Hughes S. G.
Hull D. N.
Jones D. A. C.

South N. P.
Wedgbury S. R.

FVIa/LFVIb

Carter A. A.
Cowell N. J.
Garth T. A. A.
Knights T. M.
Landmann M. J.
Chilton M. T.
Laville C. M.

FVIb

Everton G.
Ferguson H. D.
Hutchings P. W.
Kolins D. F.

Lamburn J. W.
Pemberton M. W.

FVIc

Barnes N. J.
Brightwell M. J.
Fieldhouse J.
Labrum R. C.
Leach S. D.
Oliver C. G.
Page J. W.
Smith S. P.
Ward M. J.

MVI

Bavington P. L.
Brightling J. P.

Buckley A. C.
Chau T. K-W.
Edwards P. C. R.
Hubbard T. A.
Maddox M. C. A.
Odulate O.
Pinsker M. A.
Seager-Smith P. A.
Tomkins N. G.
Woodward J. A. J.

SVIG

E Ball E. J.
Brown T. H.
Duffield W. H.
Eastoe R. L.
Frankish S. C.
Garner S. D. M.
Haigh S. A.
Hart F. P. J.
Inskip C. G. W.
Manze P. T.
Reed N. A.
Smart C. J.
Smith I. R.
Williams T. N. G.
Woodley J. M.

SVIH

Billington N. S.
Charie P. A.
Cooke S. R.
Crawford N. C.
Gray A. C.
Horton D. A.
Howe R. C.
MacLusky K. A.
Nicoll S. G.
Ognjanovic R.
Rowe P. G.
Scaglioni P. L.
Sheridan S. P.
Wilkins T. N.
Williams R. C.
Wright D. H.

EVI

Barker J. N.
Bristow G. E. F.
Caen R. D.
Caldwell R. B.
Davies M. R. N.

Hardingham T. J. P.
Hopkins J. R.
Miller P. J.
Nicholls D. C.
Robinson D. T.
Stebbings P.
Walsh C. S.
White E. J.

BVIa

Dawson C.
Dawson M. K.
Downes C. D. A.
Gulson J. D.
Hyde D. C.
Morris J. R. L.
Shepley J. M.
Surridge C.
Walbank P. C. M.
Wild M. G.

BVIb

Capon G.
Collard Q.
Collins S. R.
Cushway W.
Greenish T. S.
Griffin S. J.
Hill M. J.
Peacock M. R.
Stone A.

BVIc

Armstrong S. J. C.
Buglass J. P.
Clark Nicholas L.
Fitzsimons N.
Ibbett J. C.
Northwood M. H.
Wallis W. C.

LHVIa

Amos C. M.
Boetius A. J.
Phillips L. R.
Wilson M. I.
Workman P.

LHVIb

Chalabi S.
Dover S. J.

Fricker R. D.
Kyle T.
Lousanda T. R.
Osborne I. T.
Vidler M. J.
Wilmot P. M. B.

LHVIc

Belson N. J.
Dickinson R. W.
Jenkins D. E.
Johnson S. A.
Jupe J. R.
Mayo R. C.
Millard R. A.
Osborne M.
Scott W. S.
Stuart A. C.
Wheatley T. I.

LGVI

Allen R. J.
Charlton M. R.
Crowsley D. H.
de Vries B. van W.
Forrester S. M.
Gass G. M.
Heginbotham R. C.
Hutchinson C. G.
Illingworth S. C.
Lawrence J. C.
Workman I.

LMVI

Agboola A.
Ellwood J. A.
Kirkby O. F.
Purdy M. R.
Ratcliffe J. M.
Staddon J. A.

L SVI (Y)

Edge N. S.
Hobson R. M. G.
Howe R. R. J.
Knight R. H.
Leung W.
Lousada S. C.
Melville D. P.
Orr M. N.
Simpson P. C.
Smith P. F.

Smith T. A.
Tam P. P. Y.
Teasdel D. F.
Thiagarajah M.
Young J. W.

L SVI (G)

Ball M. G.
Buckingham P. J.
Carter C. M.
Clifton P. S.
Marson D. C.
Notley S. J.
Preston R. C. H.
Quartermaine I. R.
Sheridan R. J.
Tucker M. A. O.
Whiting C. G. A.
Williams A. G.
Woods A.

LEVI

Allingham E. de G.
Clarke G. Q.
O'Connor S.
Cook J. N.
Dawson W. T.
Greenaway N. T.
Housden R. J.
Jackson W. G.
Moore A. J.
Rayner C. S.
Salsbury M. C.
Sarwal T.
Silcock M. J.
Watson S. J.

LBVI J

Ambler R.
Caswell J. H.
Cianchi J. P.
Cockings J. G. L.
Cray S. H.
Dunkley J. D. W.
Fielden J. M.
Ghaemmaghami R.
Gurney P. D.
Inns J. H.
Lobo G. P.
Matthews I. C.
Northern S. A.
Pead M. E.

Pinney D. C.
Thomas M. J.
Toyn J. H.

LBVI P

Beazley J. C. W.
Biswell S. J.
Campbell-Gray R.
Chinn M. S.
Comins J. G. M.
Crawford R. H.
Hawkins J. R.
Johnston R. W.
Markham A. J. W.
Nixon P. P.
Parker-Eaton S. P.
Purbrick R. M.
Smith J. A.

V 2

Amos G. D.
Blake S. P.
Chilver J.
Chorley R. J.
Colbrook P. V.
Davies P. A.
Gillett-Toone R.
Green W. D.
Hall S. C.
Holes D. M.
Howe J.
Javadi A.
Kelly N. J. F.
Kingston M. P.
MacKechnie A. G.
Nicholls P. J.
Pratt M. S.
Sedding R. A. B.
Stairs A. J.
Stark R. G. K.
Taylor S. J.
Waterfield D. C. R.
Watkin P. N. W.
White J. C.

V 3

Caldwell P. J.
Cardwell A. D. D.
Charlton G. R.
Chun D. J.
Cork P. C.
Creed A. P. A.

Drozdziol P. B.
Edmonds M. M. J.
Hake C. P.
Hay J. M.
Hilton-Johnson J.
Inman M. S.
Jones R. W.
Laugharne A. D.
Lynes A. C.
Manley M. A.
Marks R. J.
Miller A. D.
Mitchel C. A.
Pearce M. W.
Robinson P. St J.
Schwimmer G.
Stevens C. A.
Sylvester J. A. C.
Thody D. M.
Welch T. C.

V 4

Allen C. E.
Baile M. T.
Bartle J. R.
Bell C. J.
Cave R.
Clayton G. N.
Coggins M. E.
Dobson J. C.
Franklin I. A.
Franks G. C.
Fulham M. R.
Hadfield G. H.
Hargreaves K. C.
Harmer W. G.
Hendry A. P.
Joiner C. P.
Kitchiner J. D.
Mence A. J.
Moser C. R. P.
Myers J. L.
Rose R. J.
Stephens D. P.
Thompson G. M.
Tsitsis A. R. W.
White M. F.

V 5

Bardner J. N.
Black W. R.
Black J. A.
Blackie C. W.

Burrell N. P.
Caffrey S.
Caves B. A.
Champion A. J.
Cook P. M.
Downes R. P.
Dart S. M.
Dunne J. A.
Freeman N. D. M.
Frost P. R.
Grant W. E. J.
Henderson D. P.
Maden P. J.
Morley S. D.
Reeken J. P.
Reid S. H.
Richardson D. C.
Rowsell C. M.
Salter A. G.
Watson M. D.
Wear J.

V 6

Allen P.
Britton I. R.
Eckert S. G.
Galley R. M.
Georgala D. M.
Ghaemmaghami A.
Grimshaw M. N.
Hadland A. J. C.
Hamilton P. C. D.
Harbinson C. T.
Henley R. D.
Hunter D. A.
Jennings P. W.
Keyworth P. C.
Lyster-Binns R. E. N.
Mostafavi M. R.
Mostafavi H. R.
Owen R. L.
Parish W. E. G.
Parrish N. M.
Penfare T. J.
Simpson N. B.
Stringer Neil A.
Sunderland M. J.

V 7

Ainsworth S.
Du Bois M. F. De G.
Hallé-Smith S. J.
Herman B. M.

Inskip L. P.
Johnson A. D.
Macpherson J. P.
Maltby H. A. J.
Moncrieff N. J. S.
Nudd L. G.
Sabey A. J.
Smith S. M.
Steed R. J.
Swift R. M.
Swift C. B.
Van Oppen S. R. M.

R 1

Barker C. J.
Baylis R. J. H.
Boner M. A.
Buckingham P. R.
Cartwright R. J.
Clements D. L.
Constable G. N. J.
Davis R. J.
Elliott H. S. M.
Harris S. J.
Kaye W. J. H.
Manze A. M.
Norris A. M.
Notley I. M.
Pull J. C. V.
Riley S. J. T.
Seager-Smith E. J.
Stambach T. A.

R 2

Adl S. M.
Aikman J. A.
Appleby M. J. C.
Cro T. E.
Dubois R. M.
Eckersley G. D.
Eltringham M. S.
Falcongreen M. F.
Grey G. McN.
Heginbotham S. D.
Horrocks T. S. G.
Hudleston N. E.
Kavan S.
Lyon S.
Maddox T. J.
Nutt M. C.'
Pinsker R. B.
Rawson P. J.
Rogers O. H.

Rowe D. S.
Russell S. L.
Turner B.
Warren S. M.
Watson J. E. S.
Woods G.

R 3

Barnes N. L. M.
Barnes A. P.
Bavington M. R.
Bell S. J.
Blane N. C.
Champion S. J.
Douglas T. M. H.
Elphee J. R.
Gifkins A. R.
Gillet B. G.
Hattan S. J.
Hopkins R. E.
Jones G. S.
Knights M. J.
Lawrence A. C.
Logan P. D.
McDonald R. J.
Morton A. T. A.
Parkinson D. S.
Richardson M. J.
Sansome C. J.
Swaney N.
Tyley J. R.
Wilks G. T. O.
Williams A. J. M.

R 4

Bluffield N. B.
Borrie N. R.
Campbell-Gray E. F.
Ching K. T.
Connell D. A.
Cordingley S. J.
Dawson G.
Dawson N. A.
Dimmock G. J.
Eales D. C.
Faulkner K. A.
Fu P. C. W.
Griffin N. E. S.
Hammond R. B.
Hartley S. J. J.
Hoare N. P.
Hughes O. J.
Kutob I. S.

Shephard D. A.
Slee C. D.
Stroud D. W. A.
Thair R. M.
Welborn R. D.
Wheatley I. M.
Wood R. J.

R 5

Briggs E. H. S.
Brodie C. J.
Burns S. I. T.
Cawthorne C. J. E.
Davies M.
Duckworth J. M.
Ewing G. I.
Grant M. P. S.
Hall A. S.
Hind J. W.
Hunt A.
Jiggle B. W.
Leeds S. C.
Masters N. K. W.
McGrath B. M.
MacLeod S. A.
Mole P.
Murray-Clarke I. G.
Pettifar T. S.
Richman J. D.
Riley P. H.
Smith M. B.
Smith G. M.
Waddington I. R.
Zaki H.

R 6

Berkeley R. F. F. M.
Birch P.
Bryant S. N.
Cartwright G. A.
Comins M. E. T.
Crow N. J.
Denton A. N.
Evans A. S.
Everett D. J.
George N. G.
Griffiths P. L.
Hemmings B. R. E.
Holbrook G. J.
Jew R. J.
Johnson T. C.
Lewis D. J. R.
Muckle P. T.

Shaw J. A.
Wilkins J.
Wrigley M. J.

R 7

Abington M. B.
Ardley J. N.
Bedford S. T. C.
Billington I. E.
Campbell-Gray T. J.
Carter T. H. D.
Davidson A. J.
Evans C. St. J.
Everett M. P.
Galvin J.
Gompertz W. E.
Kent-Baldwin P. J.
Ingle R. J.
Pell I. R.
Robson J. R. W.
Snell G. N.
Taylor A. J.

IV 1

Billing J. M.
Dunne P. S.
Greene A. S.
Hartley M. C. T.
Jones D. J.
Keech A. M.
Keer T.
Lampe D. J.
McGinty S. A.
Miller M. L. B.
Morgan R. G.
Reid S. H. M.
Shukla A.
Silcock P. J.
Squire N. S.
Tiffen P. A.
Wetherall C. J. R.

IV 2

Barrett R. P. S.
Boston N. A.
Byfield R. K.
Chavasse R.
Clark M. J. G.
Connolly A. P.
Curran F. J. MacC.
Davies E.
Gregory J. B.

Hilson J. M.
Hopkisson P. R. A.
Horton S. J.
Hubbard A. M.
Laird S. C. E.
Lyster-Binns B. E.
Mabey C. E.
Nicoll A.
Reddy J.
Tuson J. S.
Wallace A. D. G.
Wallace R. A. E.
Young I. M.

IV 3

Burch N. R. M.
Carrington J. W.
Chung S. P.
Davison R.
Flude J. C.
Gandjei R. K.
Hakim A. N.
Hallé-Smith S. C.
Heining A. P. S.
Hilton A.
Johnson M. A.
Lawrence T. J. P.
Linganayagam E.
Lousada J. D.
Martin D. A. J.
Muir A. L.
Pugh B. R.
Rance D. N. C.
Thomas S. G.
Thurston M. J.
Waterfield M. P. N.
Wilson D. N.
Ziya E.

IV 4

Booth A. C.
Burton T. G.
Cheng T. B.
Clarke V. P.
Everest G. K. R.
Fox C.
Hekmat M.
Ives M. J.
Lucas C. S.
Melville J. D.
Mitchell D. W. M.
Nickels T. B.
Panton R. A. C.

Reed A. S.
Robson R. P.
Rowe I. W.
Simpson N. C.
Smith M. A.
Thorn J. W. R.
Waldock A. A.
Welborn J. M.
White R. J.
Woodley S. A.

IV 5

Betterton M. D.
Diffey J. E.
Elliott S. S.
Fancett M. J.
Hemsley S. A.
Hendry G. R.
Ingledow R. P. J.
Jones A. M. I.
Kemble M. P.
Leung W. R.
Maslen D. J.
Miller H. K.
Pleasants R. D.
Ramage G.
Robinson S. P.
Rogers T. F. E.
Rugman A. K. M.
Stringer I. D.
Sylvester J. W. F.
Turner S. E. H.
White N.
Wilks S. P. G.
Wisbey M. D.

IV 6

Blackie A. D.
Brown J. C.
Cantle J. P.
Chapman O. V.
Chilver M.
Croot A. B. J.
Escott A. D.
Godfrey D. A.
Griffin A. C.
Harris J. A.
Hendry D. C.
Lunavat P.
Luxon D.
Noel M. J.
Parnwell A. R.
Preece D. T.

Ramage J.
Rayner K. A.
Scully S. A.
Wray J. D. L.

IV 7

Ananthanathan M.
Bardner M. R.
Barnard M. P.
Bristow B. St. J.
Bruty K. D.
Buchanan M. S.
Chorley M. D.
Davison J. P.
Donaldson C. A.
Gaishauser J. C.
Hadland D. B.
Hardwick C. E. J.
Kingston S. G.
Matthews I. A.
Rose W. M. R.
Wright G.

III1

Barlen D. N.
Bushby G. M.
Cartmell N. J.
Chilver P.
Falcongreen J. A.
Goodwin G. M. A.
Haydon G. H.
Hind M. P.
Holes W. J.
Hutchinson P. M.
Inman R. D.
Jewers C. N.
Manze S. M.
Mason D. J.
McDonald W. J.
Neely J. S.
Pitt J. M. A.
Sawyer P. S.
Smith A. E. L.
Snell L. G. C.
Sowerby P. F.
Stroud A. G. P.
Tinworth N. E.
Woodrow J. H.

III2

Albright M.
Appleby J. T. V.

Baile M. J.
Bates R. G.
Burton J. W.
Corley S. J.
Edwards M. D.
Henshaw M. F.
Illes T. M.
Leach M. P.
Li T-M. J.
Lovell D. A.
Mead R. D.
Millard J. P.
Monks S. A.
Nethersole J. P. A.
Prescott-Bran L. R.
Porter J. P.
Quartermaine M. F.
Rea C. A. L.
Robertson D. I.
Sallows M. A.
Stairs M. A.
Thelwell S. A.

III3

Allen M. E. M.
Anderson C. W.
Baber J. M.
Bromwich N.
Brown P. J.
Codrington C. E.
Cushing A. R.
Dimmock J. V.
Dite A. S. R.
Escott P. A.
Halton A. J. McA.
Holland N. L.
Hoppe C. D. L.
Kavanagh-Dowsett
S. A. H.
Nicholls I. R. J.
Parrish D. J.
Proctor M. H. N.
Saviotti P. G.
Schurink H. J. L.
Wakerly R. G.
Wakes-Miller D. H.
Watson-Lamb R. J.
Watts A. J.
Withey A. J.
Young N. J.

III4

Barker A. C. G.

	IIW	IIP	IIJ	IIH
Blackburn M. J.	Abington R. P.	Barnder T. E.	Allen S. R. A.	Allen J. D.
Bryant D. W.	Adams M. J.	Baker D. R.	Brown A. H. R.	Baldock K. W.
Campbell A. M.	Appleby R. M. W.	Dodd A. J.	Chung P. J.	Barden R. J.
Cook J. W.	Armstrong I. A.	Dyer J. R. J.	Clark E. C.	Crowe T. M.
Delaney S. R.	Ball T. H.	Flint B. J.	Cochrane A.	Cunningham M. J.
Denton M. J.	Barnard W. A.	Flude A. R.	Gazard P. D.	Dewe J.
Evans J. H.	Batho J. A.	Henley C. M.	Gibbins G. S. E.	Garner M.
Fu J. C. C.	Bavington R.	Hopkins L. D.	Gurney S. J. O.	Goodman D. W. E.
Grundon H. S.	Codd J. W.	Jones G. H. L.	Hart I. J.	James E. D.
Hart J. R.	Cross J. A. W.	Maddison J. S.	Hilson C. J.	Jones E. D. L.
Holland D. A.	Denton R. J.	Merchant S. P.	Hopkisson J. F.	Kettler J. J. R.
Johnston J. A. E.	Gentle C. G.	Nicoll C.	Kemble T. J.	Kingston S. W. R.
Moore J. J.	Groves S. R. G.	Parrott C. J.	McLeod A. J. G.	Mann D. C. E.
Muckle M. C.	Handscombe D. J.	Rawlinson A. L.	Murray A. J. H.	Moore M.
Ormerod J. P.	Hunt A. R. E.	Rayner J. A. St J.	O'Dell J.	Murthy S.
Probert D. T. G.	Ling J. S.	Shearn C. J.	Parrish J. R.	Parker-Eaton T.
Rainbow P. W.	Mercer P. J. G.	Smalls M.	Robinson A.	Rodwell P. H. W.
Rees P. R.	Murphy A. R.	Smeath R. J.	Scully P. S.	Sawford S. J.
Rogers J. A.	Ramm D. G.	Tanner I.	Smith T. G.	Stidolph R. L.
Rose G. E. J.	Sylvester J. M. B.	Taylor M. B.	Thornley K. R.	Surtees D.
Shrive J. E.	Winton G. E.	Ullman S. J.	Young R. C.	Taylor J. R. K.
Smith R. C.		Woods M.		Worthington S. M.
Wrench J. L.				

Members of Staff September 1981

T. C. Allerton
Miss S. J. Allison
E. J. Amos
J. J. G. Andre
D. A. Armstrong
D. C. Bach
C. Baker
M. E. Barlen
A. E. Barlow
C. J. Barnett
Mrs A. H. Batho
Miss L. S. Beazley
P. Boorman
P. G. Bossom
A. R. Bosworth
N. C. R. Boulting
P. C. Braggins
P. D. Briggs
D. M. Brooke
D. H. Bullock
J. W. Burgess
G. E. Caine
R. G. Caple
P. B. Churcher
K. P. Clarke
Mrs W. A. Clark
Rev. A. Clitherow
P. J. Coggins

P. E. Coles
J. I. Cooper
B. J. Cotton
N. J. Cox
B. W. Crossman
Dr A. J. Crowe
Mrs C. N. L. Crowe
Mrs J. C. Cunningham
J. H. Davidson
Rev. P. R. Davies
R. A. Eadie
T. J. Elliott
Mrs E. J. Elliott
F. M. Fletcher
Mrs F. M. Fletcher
G. M. K. Fletcher
J. R. Fletcher
J. L. Foulkes
J. L. Freame
R. F. Goodacre
P. V. Guy
S. D. Hale
R. C. Hadley
K. S. Hall
G. S. Haynes
Rev. M. D. A. Hepworth
R. J. Hudson
D. B. T. Hughes

Miss S. M. Hughes
Mrs J. R. Hull
D. W. Jarrett
C. G. Jeavons
R. K. Johnson
G. G. J. Jones
R. E. R. Jones
N. J. Keatley
Rev. J. C. Laird
I. Lawrence
P. M. MacMahon
T. J. Machin
Maj. D. R. Mantell
M. J. Marks
C. L. Marsh
H. F. McKendrick
A. R. Millett
Dr G. W. Mines
D. F. Moore
A. W. Morris
M. J. Norris
D. S. Odom
Mrs L. J. O'Hara
J. B. D. Osborne
J. C. Osman
K. Parkinson
S. J. Partridge
G. N. Phillips

F. J. Pike
G. F. W. Pleuger
D. C. Poulton
D. B. Prior
P. F. Ramage
K. B. Rapson
M. J. Rawlinson
A. J. Reid
T. A. Riseborough
I. J. Rosser
I. Schofield
D. K. Sewell
Mrs N. B. Smith
M. P. Stambach
C. J. Steen
D. P. C. Stileman
J. Sylvester
A. M. Thorp
G. D'E. Trevelyan
K. S. Tromans
M. R. Vogel
D. Warren
Miss C. F. Watson
G. J. Wickens
P. S. Wiser
P. R. O. Wood
P. A. Young

Names of Boys by forms Christmas Term 1981

VI

Ainsworth S. R.
Eckert S. G.
Kaye W. J. H.
Manze A. M.
Norris A. M.
Swift R. M.
Welch T. C.

FVIa

Cave R.
Coggins M. E.
Drozdziol P. B.
Freeman N. D. M.

Gillett-Toone R.
Mackechnie A. G.
Sedding R. A. B.
Tsitsis A. R. W.

FVIb

Hilton-Johnson J.
Otten G. V.

HVI@

Amos C. M.
Bailey M. B.
Phillips L. R.

EngVI@

Mayo R. C.

HVIa

Blake S. P.
Boetius A. J.
Chilver J.
Downes R. P.
Grant W. E. J.
Green W. D.
Hadfield G. I. H.
Hendry A. P.
Pratt M. S.
Rose R. J.

Sylvester J. A. C.
Wilson M. I.

HVIb

Allen C. E.
Bell C. J.
Cartwright R. J.
Constable G. N. J.
duBois M. F. De G.
Galley R. M.
Hoskins R.
Jennings P. W.
Keyworth P. C.
Reid S. H.

HVIc

Amos G. D.
Dobson J. C.
Elliott H. S. M.
Henderson D. P.
Morley S. D.
Schwimmer G.

MVIa

Chan R. T. C.
Charlton G. R.
Clements D. L.
Kingston M. P.
Mence A. J.
Stephens D. P.

MVIb

Bardner J. N.
Dunne J. A.
Georgala D. M.
Henley R. D.
Owen R. L.
Thody D. M.

SVI@

Carter C. M.
Notley S. J.
Williams A. G.

SVIC

Bhagat R.
Black W. R.
Blackie C. W.
Boner M. A.
Bromley J. R.
Caffrey S.
Champion A. J.
Chan J. C. C.
Hassanein S.
Holes D. M.
Howe R. J.
Javadi A.
Kazi A. N.
Maden P. J.
Miller A. D.
Mostafavi H. R.
Nicholls P. J.
Stark R. G. K.
Stringer N. A.
Watkin P. N. W.
White M. F.
Wotherspoon K. R.

SVIH

Allen P.
Buckingham P. R.
Chorley R. J.
Chun D. J.
Davis R. J.
Edmonds M. M. J.
Frost P. R.
Hamilton P. C. D.
Harbinson C. T.
Harris S. J.
Ihonor C. C.
Jones R. W.
Kelly N. J. F.
Liu S. C. P.

Marks R. J.
Mostafavi M. R.
Notley I. M.
Nudd L. G.
Pull J. C. V.
Seager-Smith E. J.
Stairs A. J.
Wear J.

GVI

Barnard S. T.
Hunter D. A.
Maltby H. A. J.
Richardson D. C.
Riley S. J. T.
Salter A. G.
Taylor S. J.

EVI@

Silcock M. J.

EVI

Caves B. A.
Cook P. M.
Hadland A. J. C.
Halle-Smith S. J.
Hargreaves K. C. McA.
Inman M. S.
Johnson A. D.
Löhner C.
Myers J. L.
O'Brien H. E.
Waterfield D. C. R.

BVI@

Fielden J. M.
Toyn J. H.

BVIa

Baylis R. J. H.
Colbrook P. V.
Davies P. A.
Hall S. C.
Hay J. M.
Lobo G. P.
Lynes A. C.
Matthews I. C.
Pinney D. C.
Stambach T. A.
Stevens C. A.
Thompson G. M.
White J. C.

BVIb

Bartle J. R.
Clayton G. N.
Fraser S. G.
Hake C. P.
Manley M. A.
Mitchel C. A.
Parish W. E. G.
Penfare T. J.
Sunderland M. J.
Watermeyer S. R.

BVIc

Baile M. T.
Black J. A.
Caldwell P. J.
Franks G. C.
Fulham M. R.
Grimshaw M. N.
Singh M.

LVI

Hartley M. C. T.
Wetherall C. J. R.

LFVIa

Dawson N. A.
Dubois R. M.
McCracken Niall
Williams A. J. M.

LFVIb

Elphee J. R.
Hopkins R. E.
Parkinson D.
Thair R. M.
Tyley J. R.

LHVIa

Aikman J. A.
Blane N. C.
Eckersley G. D.
Eltringham M. S.
Heginbotham S. D.
Morgan R. G.
Nutt M. C.
Rogers O. H.
Russell S. L.
Squire N. S.
Watson J. E. S.

LHVIb

Crow N. J.
Everett D. J.
Galvin J.
Hammond R. B.
Henry P. B.
Muckle P. T.
Pettifar T. S.
Shephard D. A.
Stevens J. C.
Van Oppen S. R. M.
Wilks G. T. O.

LHVIc

Dawson G.
Dimmock G. J.
Douglas T. M. H.
Eales D. C.
Evans C. St. J.
Everett M. P.
Gompertz W. E.
Hunt A.
McGrath B. M.
Roach R.

LMVIa

Appleby M. J. C.
Davies M.
Keech A. M.
Keer T. J.
Pinsker R. B.
Silcock P. J.

LMVIb

Champion S. J.
Falcongreen M. F.
Ingle R. J.
Lyon S.
Welborn R. D.

LMecVI

Brodie C. J.
Hall A. S.
Lawrence A. C.
McDade E. K.

LSVIY

Bavington M. R.
Bell S. J.
Billing J. M.
Borrie N. R.

Cro T. E.
Duckworth J. M.
Dunne P. S.
Gillett B. G.
Greene A. S.
Johnson T. C.
Jones D. J.
Lampe D. J.
McDonald R. J.
Miller M. L. B.
Pickersgill M. R.
Reid S. H. M.
Shukla A.

LSVIC

Berkeley R. F. F. M.
Birch P.
Comins M. E. T.
Ewing G. I.
Faulkner K. A.
Fu P. C. W.
Gifkins A. R.
Hughes O. J.
Jones G. S.
Kavan S.
Knights M. J.
McGinty S. A.
Mok H. L. H.
Richardson M. J.
Rowe D. S.
Suen D. T. M.
Swaney N.
Voelcker R. M.
Wood R. J.

LEVI

Denton A. N.
Evans A. S.
George N. G.
Hind J. W.
Kutob I. S.
Maddox T. J.
Shaw J. A.
Smith G. M.
Stroud D. W. A.

LBVIa

Adl S. M.
Barker C. J.
Barnes N. L. M.
Ching K. T.
Cordingley S. J.

Grey G. McN.
Horrocks T. S. G.
Hudleston N. E.
Swan C. S. G.
Wheatley I. M.
Riley P. H.
Hatton S.

LBVIb

Barnes A. P.
Bryant S. N.
Burns S. I. T.
Cartwright G. A.
Hemmings B. R. E.
Holbrook G. J.
Lewis D. J. R.
Robson J. R. W.
Sansome C. J.
Slee C. D.
Snell G. N.
Zaki H.

LBVIc

Billington I. E.
Campbell-Gray T. J.
Campbell-Gray E. F.
Carter T. H. D.
Howard J. C.
Jiggle B. W.
McLeod S. A.
Wrigley M. J..

V2

Barrett R. P. S.
Boston N. A.
Braun M.
Chavasse R.
Clark M. J. G.
Connolly A. P.
Curran F. J. McC.
Davies E.
Flude J. C.
Gandjei R. K.
Gregory J. B.
Hilson J. M.
Hilton A. I.
Hopkisson P. R. A.
Horton S. J.
Laird S. C. E.
Lyster-Binns B.
Mabey C. E.
Nicoll A.

Purdy S. P.
Reddy J.
Tiffen P. A.
Tuson J. S.
Waldock A.
Young I. M.

V3

Carrington J. W.
Daniel S. F.
Davison R.
Griffin N. E. S.
Hakim A. N.
Halle-Smith S. C.
Heining A. P. S.
Hekmat M.
Hubbard A. M.
Johnson M. A.
Lawrence T. J. P.
Linganayagam E.
Lousada J. D.
Martin D. A. J.
Muir A. L.
Pugh B. R.
Rance D. N. C.
Tear M. D.
Thomas S. G.
Thurston M. J.
Wallace A. D. G.
Wallace R. A. E.
Waterfield M. P. N.
Wilson D. N.
Ziya E.

V4

Booth A. C.
Burton T. G.
Cheng T-w B.
Clarke V. P.
Everest G. K. R.
Fox C.
Hay A. D. S.
Hodson N. B.
Ives M. J.
Lucas C. S.
Melville J. D.
Mitchell D. W. M.
Nickels T. B.
Panton R. A. C.
Rayner K. A.
Reed A. S.
Robson R. P.
Rowe I. W.

Simpson N. C.
Smith M. A.
Thorn J. W. R.
Welborn J. M.
White R. J.
Wilks S. P. G.
Woodley S. A.

V5

Betterton M. D.
Cantle J. P.
Diffey J. E.
Elliott S. S.
Fancett M. J.
Harris J. A.
Hemsley S. A.
Hendry G. R.
Ingledow R. P. J.
Jones A. M. I.
Kemble M. P.
Leung W-k. R.
Maslen D. J.
Miller H. K.
Pleasants R. D.
Ramage I. G.
Robinson S. P.
Rogers T. F. E.
Rugman A. K. M.
Stringer I. D.
Sylvester J. W. F.
Turner S. E. H.
White N.
Wisbey M. D.
Wright G. R.

V6

Blackie A. D.
Brown J. C.
Chapman O. V.
Chilver M.
Croot A. B. J.
Davidson A. J.
Donaldson C. A.
Escott A. D.
Godfrey D. A.
Griffin A. C.
Hendry D. C.
Luxon D.
Lyster-Binns R. E. N.
Matthews I. A.
Myers P. A.
Noel M. J.
Parnwell A. R.

Preece D. T.
Ramage J.
Scully S. A.
Taylor A. J.
Wilkins J.
Wray J. D. L.

V7

Abington M. B.
Ananthanathan M.
Ardley J. N.
Bardner M. R.
Barnard M. P.
Bedford S. T. C.
Bristow B. J. St J.
Bruty K. D.
Davison J. P.
Fish D. A.
Gaishauser J. K.
Hadland D. B.
Hardwick C. E. J.
Herd S.
Kingston S. G.
Odulate A.
Rose W. M. R.
Smith S. M.
Tomlinson J. F.
Welsh C. J.

R1

Barlen D. N.
Buckland M. A.
Bushby G. M.
Chilver P.
Haydon G. H.
Lawson-Smith A. E. L.
McDonald W. J.
Manze S. M.
Neely J. S.
Pitt J. M. A.
Prentice I. R.
Sawyer P. S.
Tinworth N. E.
Woodrow J. H.

R2

Albright M. S.
Beckwith T. J.
Carruthers A. S.
Cartmell N. J.
Castenskiold E. H.
Chapman R. P.

Crankshaw D. J.
Dover R. T. A.
Falcongreen J. A.
Hewlett M. J.
Hind M. P.
Holes W. J.
Hutchinson P. M.
Inman R. D.
Jewers C. N.
Lovell D. A.
McAra D. R.
Normoyle M.
Prescott-Brann L. R.
Saxby C. W.
Sowerby P. F.
Stroud A. G. P.
Watson-Prain N. H.
Webb M. K.

R3

Adl C.
Anderson C. W.
Appleby J. T. V.
Bates R. G.
Burton J. W.
Corley S. J.
Giddings A. C.
Goodacre I. R.
Goodwin G. M. A.
Henshaw M. F.
Iles T. M.
Leach M. P.
Mead R. D.
Millard J. P.
Monks S. A.
Nethersole J. P. A.
Rea C. A. L.
Snell L. G. C.
Stairs M. A.
Stanley S. J.
Thelwell S. A.
Whitaker L. K.
Willis W. R.

R4

Baile M. J.
Bayfield M. C.
Brown P. J.
Claxton G. E.
Codrington C. E.
Compton A. J.
Cranwell P.
Dudley J. C.

Edwards M. D.
Ellingham S. P.
Kavanagh-Dowsett S. A. H.
Lavender G. W.
Lloyd Dieter S.
Marshall B. H. J.
Muckle M. C.
Pendred C. R.
Place D.
Porter J. P.
Proctor M. H. N.
Robertson D. I.
Schurink H. J. L.
Wakerly R. G.
Williams R. W.
Young N. J.

R5

Allen M. E. M.
Bromwich N. P.
Campbell A. M.
Chapman N. J. A.
Dawson R. D.
Denton M. J.
Escott P. A.
Framp K. P.
Fu C. C. J.
Halton A. J. McA.
Holland N. L.
Hoppe C. D. L.
Ireland C. D.
Moore J. James
Nicholls I. R. J.
Pattison M. A. I.
Probert D. T. G.
Quartermaine M. F.
Rees P. R.
Rogers A. M.
Saviotti P. G.
Smith A. R. J.
Stevens M. R.
Tan E. L. T.
Thody D. A.

R6

Barber J. M.
Barker A. C. G.
Dite A. S. R.
Downie J. A.
Hardwick N. P. A.
Hart J. R.
Holland D. A.

Ingle A. F.
Johnston J. A. E.
Joy A. J.
Martyn M. J. G.
Must R. C.
Parrish D. J.
Portsmouth S. D.
Wakes-Miller D. H.
Watson-Lamb R. J.
Watts A. J.
Winton T. R.
Withey A. J.
Wrench J. L.
Wright J. N.

R7

Blackburn M. J.
Bryant D. W.
Buchanan G. A.
Cook J. W.
Cushing A. R.
Dawson A. C.
Delaney S. R.
Evans J. H.
Grundon H. S.
Ormerod J. P.
Rainbow P. W.
Rogers J. A.
Rose G. E. J.
Shrive J. E.
Smith R. C.
Walker D.

IV1

Barker D. R.
Cochrane A.
Dyer J. R. J.
Fenton J. A. L.
Gardner R.
Gurney S. J. O.
Henley C. M.
Herbert P. J. W.
Hope T.
Hopkins L. D.
Jones G. H. L.
Mitchell P. G.
Nicoll C.
Smith T. G.
Taylor M. B.
Thornley K. R.
Ullman S. J.

IV2

Adams J. P.
Allen S. R. A.
Barnard W. A.
Brown A. H. R.
Cunningham M. J.
Flint B. J.
Gibbins G. S. E.
Greetham T.
Hay A. C. W.
Jones E. D. L.
Kemble T. J.
Kenyon P. R.
Legg C. J. G.
Maddison J. S.
Merchant S. P.
Murray A. J. H.
Pickard T. E.
Scully P. S.
Smeath R. J.
Von Winterfeldt C. C.
Wetherall R. C. R.
Wharton R. Q.

IV3

Barker J. J. P.
Codd J. W.
Cotton M. H.
Davison P. M. A.
Flude A. R.
Gifkins M. P.
Gregory N. P.
Hart I. J.
Hilson C. J.
Hopkisson J. F.
Howard J. E.
James E. D.
Murnal P.
Murphy A. R.
O'Dell J. S.
Rawlinson A. L.
Robinson J. A.
Sansome A. D.
Shearn C. A.
Smalls M.
Spencer D. E. G.
Surtees D. W.
Young R. C.

IV4

Armstrong I. A.
Barden R. J.
Bardner T. E.
Burgess N. J.
Caffrey J.
Coley J. C.
Crooker J. E.
Dewe J.
Dimmock J. V.
Dunkley C.
Harper R. J.
Hudson D. D.
Ling J. S. G.
McLean P. J.
Marshall J. A. L.
Measham C. J.
Murthy S.
Parrish J. R.
Ramm D. G.
Robinson A. P. C.
Wills D. J.
Winton G. E.
Woods J.

IV5

Abington R. P.
Baldock K. W.
Bell N. A.
Charlton A. S.
Cox S. J.
Cross J. A. W.
Doubleday J. J.
Fawcett J. M.
Geary J. D.
Grosvenor A. E.
Groves S. R. G.
Hunt A. R. E.
Husband S. P.
Lam G. Y. S.
Lyons T. N.
McLeod A. J. G.
Mercer P. J. G.
Parrott C. J.
Stidolph R. L.
Stillwell N. J.
Sylvester J. M. B.
Tanner I.
Worthington S. M.

IV6

Appleby R. M. W.

IV4

Ball T. H.
Batho J. A.
Bavington R. F.
Charsley N. A.
Davies R. T.
Garner M.
Goodman D. W. E.
Kennedy D.
Kingston S. W. R.
Mann D. C. E.
Moore E. J.
Moore M.
Putterill M. C.
Rodwell P. H. W.
Smalley C. D.
Vacy-Ash J. N.

IV7

Adams M. J.
Allen J. D.
Ball M. P. S.
Collett M. R. D.
Davidson J. C.
Denton R. J.
Gentle C. G.
Hale R. D. A.
Handscombe D. J.
Kettler J. J. R.
Marshall A. S.
Parker-Eaton T. S.
Rutherford D. G.
Sawford S. J.
Stanley G.
Taylor J. R. K.
Thompson S. R. J.

III1

Boulting N. E.
Briscoe J. D.
Cockings G. F. S.
Davies R. A.
Derbyshire D. A.
Eltringham S. D.
Freeman H. Lucas
Goodacre C. R.
Goodwin T. J. L.
Hill D. A.
Holborn N. L.
Issitt C. A.
Makela J.
Martin J. C.
Martin S. B.

Merchant D. J.
Miller G. A.
Nicholas O. M.
Pebody J. E.
Rees N. D.
Stanley A. G.
Taylor D. P. K.
Tongue A. G.
Ullman C. G.
Woodbridge A. J.
Woodrow N. C.

III2

Banks B. C.
Booth R. J.
Briggs S. M.
Carpenter P. D.
Cartmell A. B.
Caves M. J.
Dubois A. J.
Fish R. A.
Hall M. J.
Howard G. E. O.
Irish B. D. R.
Jewers M. O.
Johnson R. J.
Leaver M. H. O.
McCracken S. F.
Maynard J. P.
Must N. H.
O'Dell D. A.
Riding R. D.
Savva G.
Sawyer T. M.
Scott A. L.
Sims G.
Smedley M. J.
Staddon D. J.
Young D. W.

III3

Allerton D. C.
Barnett H. P.
Cumberland S. R.
Cummings D. P.
Devenish D. M.
Donald I. W.
Foster I. E.
Hammond J. M. P.
Harvey R. E.
Hotchkies B. J.
James E. A.

Lewis J. C.
Milne A. J.
Mitchell R. G.
Mooring R. J.
Norman R. L. A.
Payne R. J.
Rimmer J. D.
Rutgers K. T. F.
Sexton M. J.
Smith A. F.
Sylvester J. B.
White J. C.
Willsher A.
Winton J. A. C.

III4

Bates A. C. D.
Berrecloth B. J.
Bruty G. B. M.
Cameron E. J.
Case W. D.
Caves P. J.
Coladangelo C. A.
Currie P. J. J.
Evans M. W.
Ferdinando J. S.
Halle-Smith N.
Hammond J. P.
Hoole A. J.
Kettler C. C. R.
Kitchiner A. D.
MacLeod J. K.
McGruer J. B.
Minor G. P. L.
Parrish C. E.
Rasche C.
Sturges P. E.
Taylor S. B.
Watson R. J.
Woodford E. C.
Wootton T. E.

IIH

Armstrong Q. D.
Bowen G. M.
Cannings S. J.
Cave M. R.
Fielding P. A.
Goodman J. G. E.
Gordon J. E.
Green J. A.
Greenslade D. J.

Hale E. J.
Jones P.
Kemble P. J.
Landmann C. S.
Ling T. J. M.
McLeod R. C. T.
Marshall A. C.
Parsons L. N.
Peacock S. C.
Podmore M. D. W.
Rasche T.
Sexton S. R.
Shearn J. G. J.
Taylor P. A. W. W.
Wilks A. F.

IIW

Bates E. P.
Bedrich M. G. C.
Bradley G. J.
Collins D.
Costello S. W.
Crankshaw A. R.
Cranwell M. A.
Dempster I. M.
Falcongreen D. S.
Francis R. J.
Henman A. N.
Hine P. W. J.
Jew N. J.
Millman S. G.
Morris R. P.
O'dell W. T.
Owen W. D. T.
Pell C. D.
Rhodes J. T. E.
Timson C.
Tollman M. O.
Walbank M. H.
Williams J. H. P.
Wood M. D.

IIP

Bardner B. W.
Brodie J. R.
Coleman J. R. B.
Cutress R. I.
Drakard M. R.
Gamble M. J.
Harris J. K.
Jack M. J. E. S.
Jones R. A.

Jones J. A. D.
Lilley J. G.
Lopez-Real C. E.
Perl J. R.
Phipps S. J.
Quinn S. D.
Ramage N.
Raynor P. D.
Rees G. D.
Sales J. R.
Shuba F.
Smith J. W.
Truelove J. C. S.
Veitch S. A.
Wakes-Miller H. J.

IIJ

Armstrong K. J.
Parker D. P.
Blackley A. S.
Brown T. M.
Brown J. D.
Buchanan G. N.
Burley N.
Burton D. B.
Dingley G. J. C.
Dodd J. K. F.
Fields J. R.
Grove P.
Hall R. A.
Harling N. A.
Hope A. E.
Jones T. D. W.
Lander R.
Lersch S. W. A.
Machin G. L.
McDonald M. V.
Mukherjee U. B.
Rimmer D. J.
Thatcher J. M.

Preparatory School

Staff 1979		Staff 1981	
Miss A. S. Awdry	Miss S. M. Hughes	Mrs C. N. L. Crowe	P. M. MacMahon, Esq.
Miss C. F. Carter	P. M. MacMahon, Esq.	Mrs J. Davidson	A. R. Millett, Esq.
Mrs C. N. L. Crowe	A. R. Millett, Esq.	Mrs E. J. Elliott	D. C. Poulton, Esq.
Miss S. U. J. Findlay	D. C. Poulton, Esq.	G. M. K. Fletcher, Esq.	K. S. Tromans, Esq.
G. M. K. Fletcher, Esq.	Mrs C. Powell	R. C. Hadley, Esq.	Miss C. F. Watson
R. C. Hadley, Esq.		Miss S. M. Hughes	

Headmaster P. F. Ramage, Esq.

Boys in the Preparatory School 1981 (by House)

Armstrong Kristian J.
Bardner Benjamin W.
Barker David P.
Brodie James R.
Brown Trevor M.
Brown Julian D.
Buchanan Gordon N.
Burton Daniel B.
Cannings Simon J.
Cave Matthew R.
Dingley Guy J. C.
Fields Julian R.
Gazard Daniel J.
Green John A.
Hale Edward J.
Hall Robin A.
Harling Nicholas A.
Jones Theodore D. W.
Lander Rupert
Ling Timothy J. M.
Lopez-Real Carlos E.
Machin Gareth L.
McDonald Michael V.
Mukherjee Uday Bhanu
Perl Jeremy R.
Sales John R.
Thatcher John M.
Wakes-Miller Hamish J.

Bates Edward P.
Bowen Gareth M.
Coleman Jonathan R.B.
Cutress Ramsey I.
Davies Craig N.
Dodd Jeremy K.F.
Drakard Matthew R.
Falcongreen Damon S.
Fielding Patrick A.
Gordon Jonathan E.
Grove Patrick
Henman Andrew N.
Jew Nicholas J.
Jones Jonathon A. D.
Kemble Peter J.
Landmann Colin S.
Lilley Jonathan G.
McLeod Robin C. T.
Owen William D. T.
Raynor Paul D.
Rhodes, James T. E.
Sexton Simon R.
Shearn James G.J.

Walbank Michael H.
Williams John H. P.
Wood Moray D.

Armstrong Quintin D.
Bedrich Mark G. C.
Bradley Graham J.
Costello Sean W.
Crankshaw Andrew R.
Dempster Iain M.
Francis Richard J.
Goodman James G. E.
Hine Phillip W. J.
Jones Robert A.
Jones Paul
Millman Stephen G.
O'Dell W. Toby
Pell Christopher D.
Podmore Matthew D. W.
Ramage Nicholas
Rasche Thomas
Rees Graham D.
Sear Christopher J.
Tollman Mark O.
Truelove James C. S.
Wilmot Sean W.
Wilks Andrew F.

Austoni Piero M.
Borrie A. J. Stuart
Brough Jonathan
Bygrave Michael L.
Chadwick Simon C. J.
Crotty Simon M. P.
Cummings Julian T.
Eley Jonathan E.
Hameed Suhail
Hill Christopher T. E.
Kilroy Jonathan C. R.
Lake Gareth S.
Lavis Matthew B.
Lopez-Real Dominic J.
Manning Joseph J.
Pebody Roger M. T.
Peer Thomas H.
Pinkney David S. McC.
Poulter Robert A.
Prosser David N.
Smeath Alexander M.
Wheatley Justin J.
Yovichic David

Allen Richard J.
Atkins Graeme C.L.
Attenborough Royd J.
Capon Neal R. J.
Carpenter Ian
Cartmell Cameron G.
Clark Thomas A.
Hale Philip M.
Harris Adam G.
Heslett Mark W. J.
Hodgson John A. C.
Hulbert Jonathan
Jones Robert H.
Khiani Raj M.
Logan Paul D.
Martin William H.
Miller Robert A.
Riley Robert H. G.
Sarsfield Willian P. T.
Simmons Christopher J.
Slack Mark J.
White James P.
Witham Karl E.

Abraham Dean J.
Broderick Alistair J.
Brown Howard S.R.
Buckle Marcus A.
Carpenter Christopher J.
Clark James B. D.
Constable Piers P. J.
Dickins Richard C. E.
Green Jeremy M. A.
Irons Richard
Maynard George F.
Meadwell Simon A.
Millar Paul
Pleasants J. Justin
Saunders Piers B.
Simpson Robert
Truelove Michael J.
Walsingham Rowland N.
Watson Justin J.
Young Matthew J.

Bone Timothy J.
Brown Alec T.
Daker George H.
Dunne Alexander D.
Falcongreen Lars C.
Farnworth Oliver G.
Grieve James S. L.

Hale Andrew
Hartwell Andrew
Khiani Anil
Lippmann Simon F.
MacLennan William A.
Martin Dominic W.
Miller John C.
Morris Jocelyn C. L.
Morris John W.
Parsons Nicholas S. A.
Pilgrim Jonathan C.
Quinn Paul A.
Rodgers Neil E.
Semark David E.
Shrive Dorian C.
Taylor Giles E. A.
Turner Douglas J. B.

Eastcott Daniel N.
Gutteridge Samuel C. D.
Harris Neil
Haslem Nicholas J. De L.
Heeley Stuart B.
Izzard Matthew J.
Lake Mark G.
Langdon Toby B. J.
Mallett Carl W.
Michalski Andrzej R.
Pape Michael B.
Prudence John N.
Riding Charles R.
Rose James E.
Savva Andrew P.
Tinkler Oliver T.
Tollman Nigel J.
Veitch Paul J.
Wong Chao Kang

Brewer James N.
Bryant Darren P.
Bull Raoul H.
Dring Philip W. J.
Gentle Christian H. R.
Gooch Jonathan D.
Kettler Richard H. R.
Moore Fergus G.
Morris Henry T.
Parrish William K.
Ross Neil T. MacD.
Scott Thomas Alexander
Shaw Andrew H.
Wiltshire Mark J.

Ambrose Duncan deM.
Barber Alan J.R.
Bath Thomas G.
Betts Jacob T.
Bryant Matthew C.
Bygraves Joel A.
Foster Christopher T.
Fullerton Adrian P.L.G.
Garman Jeremy D.
Haslem Dominic C. De L.
Hepworth Jeremy D.
Hill Gregory A.
Holton Stuart B. A.
Juffs Alexander W.
Kettleborough Simon
Ledsom Mark
Martin George M.
Miller Ben J. A.
Momen Sedrhat
Neale Alexander C.
Nevard David P. R.
Parkinson Timothy J.
Richardson Garrie E.
Shirley Andrew R. H.
Williams Kevin A.
Wilmot Robert

166